PROGRESS IN
MATHEMATICS

Volume 13

Probability Theory, Mathematical Statistics,
and Theoretical Cybernetics

PROGRESS IN MATHEMATICS
Translations of Itogi Nauki — Seriya Matematika

PROGRESS IN
MATHEMATICS

Volume 13

Probability Theory, Mathematical Statistics, and Theoretical Cybernetics

Edited by
R. V. Gamkrelidze

V. A. Steklov Mathematics Institute
Academy of Sciences of the USSR, Moscow

Translated from Russian by J. S. Wood

ℙ PLENUM PRESS · NEW YORK–LONDON · 1972

The original Russian text was published for the All-Union Institute of
Scientific and Technical Information in Moscow in 1970 as a volume of
Itogi Nauki — Seriya Matematika

Library of Congress Catalog Card Number 67-27902

ISBN 978-1-4684-8081-8   ISBN 978-1-4684-8079-5 (eBook)
DOI 10.1007/978-1-4684-8079-5
The present translation is published under an agreement with
Mezhdunarodnaya Kniga, the Soviet book export agency

© 1972 Plenum Press, New York
Softcover reprint of the hardcover 1st edition 1972

A Division of Plenum Publishing Corporation
227 West 17th Street, New York, N.Y. 10011

United Kingdom edition published by Plenum Press, London
A Division of Plenum Publishing Company, Ltd.
Davis House (4th Floor), 8 Scrubs Lane, Harlesden, NW10 6SE, London, England

Preface

This work is a continuation of earlier volumes under the heading "Probability Theory, Mathematical Statistics, and Theoretical Cybernetics," published as part of the "Itogi Nauki" series.

The present volume comprises a single review article, entitled "Reliability of Discrete Systems," covering material published mainly in the last six to eight years and abstracted in "Referativnyi Zhurnal — Matematika" (Soviet Abstract Journal in Mathematics). The bibliography encompasses 313 items.

The editors welcome inquiries regarding the present volume or the format and content of future volumes of the series; correspondence should be sent to the following address: Otdel Matematika (Mathematics Section), Baltiiskaya ul., 14, Moscow, A-219.

Contents

RELIABILITY OF DISCRETE SYSTEMS
M. A. Gavrilov, V. M. Ostianu, and A. I. Potekhin

Reliability of Discrete Systems

M. A. Gavrilov, V. H. Ostianu, and A. I. Potekhin

Introduction

Four years have elapsed since the publication of [19], in which the mathematical aspects of reliability theory were surveyed. In the interim a great many papers have been written on the subject, making the theory of discrete system reliability a difficult area to survey in total perspective. The special problem of the improvement of discrete system reliability by the application of structural redundancy has grown into a full-fledged discipline, namely the problem of synthesizing discrete systems with a structure designed to abate the effects of component failures and unequal delays in the activation of those components on overall system operation. Consistent with the Soviet terminological state standard (GOST) [140], reliability is defined as the capability of an object to execute prescribed functions while maintaining its operational specifications within assigned limits for a required time period or during the performance of a required task. The reliability of discrete systems actually has two interpretations:

a) "infallibility,"* in the sense of the capacity of a discrete system to execute prescribed functions (the prescribed functional

---
*Translator's Note: These distinctions are more precise than is customary in the literature surveyed. Thus, in the interest of consistency with the latter, the term "reliability" is used throughout Chap. I in the narrow sense defined above, i.e., "infallibility" (also called "spatial reliability," "equipment reliability," "failure tolerance," etc.). Elsewhere (Chap. III in particular) "reliability" has the more generic sense of over-all (space-time) reliability, i.e., "infallibility" plus "stability." The latter term, which occurs primarily in Chap. II, is sometimes called "temporal reliability."

algorithm*) in the event of failure of a certain number of system components;

b) "stability,"† in the sense of the capacity of a discrete system to execute a prescribed functional algorithm in the event of changes in the time parameters of the system.

More than 600 papers have been published to date on the topic of discrete system reliability. They may be grouped into the following categories:

1) the analysis of reliability and calculation of the probabilistic characteristics of reliability (efficiency, service life, repairability);

2) the assurance of operational reliability (renewal theory, preventive inspection, maintenance problems, etc.);

3) methods for the synthesis of so-called "reliable" structures, i.e., the synthesis of stable structures whose functional reliability characteristics are improved through the use of redundancy.

The reliability of discrete control devices is assured not only by the selection of high-reliability components, but also by the introduction of redundancy, which makes it possible to achieve a desired level of "infallible" and "stable" operation even with unreliable components. This problem is the primary concern of the present article.

In the article we not only discuss problems that have already been solved, but also attempt to give a general view of the state of the art of various models for the introduction of redundancy into the structure of discrete systems, to organize a classification of those models, and to portray all significant research trends with regard to the problem. In order to compact the text we exclude any discussion of the problems involving computation of the degree of reliability of discrete systems, despite their intimate bearing on the improvement of reliability, because they constitute an independent topic and rely on a wholly different mathematical approach and methodology.‡

---

*See [142] for the precise definition of the functional algorithm and other [Russian] terms in the theory of relay devices.

†See note on preceding page.

‡Detailed surveys of the problems may be found in [32, 33, 107, 4, 20, 172, 87, 67, 183].

The article is divided into three chapters in accordance with the major problems outlined above. In the first chapter we discuss the problem of investing systems with higher operational reliability through abatement of the influence of component failures on over-all system operation. In the second chapter we investigate the enhancement of the operational "stability" of a system through elimination of the influence of so-called "hazards" and "race conditions," i.e., through the elimination of inopportune changes of state of the system components. In the third chapter we analyze the combined problem of enhancing the operational reliability and stability of discrete systems.

CHAPTER I

ASSURANCE OF INFALLIBILITY
IN DISCRETE SYSTEMS

§ 1. State of the Art

The earliest studies of reliability assurance by the introduction of redundancy were basically concerned with the notion of simple composition (replication). Composition methods are time-honored and merely entail the complementation of a particular component or module of a system with one or more redundant replicas of the component or module, which either take over the function of the primary object in the event of failure (passive, or "cold," composition) or operate simultaneously with the primary object (active, or "hot" composition). Essentially, however, the modern theory of reliability assurance through the introduction of redundancy was developed and its fundamental disciplines defined at a later date. This theory had its genesis in von Neumann's six lectures on probabilistic logic at the California Institute of Technology in 1952; the lectures were subsequently published in the now-classical treatise [269]. Somewhat later Moore and Shannon published their paper [263] on the synthesis of reliability circuits from unreliable relays. With the advent of information theory, particularly the branch thereof known as coding theory, the growing tendency was to regard discrete systems as special kinds of channels, namely computation channels, and to treat them with the reliability improvement techniques used in communication channels. The legitimacy of this approach was investigated by Elias [196], and the practical application of coding to discrete systems of a particular type was explored by Gavrilov [9] and Zakrevskii [35].

More than 300 papers have appeared in the last decade dealing with various aspects of redundancy as applied to the enhancement of discrete system reliability. In its complete form, however, i.e., with regard for the demands imposed on system structures, minimization of redundancy in the structure or inputs, the simultaneous asset of simplicity, selection of the most effective codes, etc., the problem of reliability improvement through redundancy has yet to be solved. Most of the papers published to date merely treat isolated aspects of the problem and arrive at *ad hoc* solutions or survey special problems [12, 17, 297, 285, 245, 166, 227, 114]. Nevertheless, even the solution of these specialized problems requires a certain idealization with respect to the construction of various models to approximate in some degree the actual properties of discrete control devices or to delineate the construction and function of individual "blocks" and their interrelationships. Due to the complexity of the statement of the redundancy problem in the large there is a total lack of papers broad enough in scope to permit some assessment of the optimum model to meet a specific set of conditions. The survey that follows covers the existing literature with regard to individual areas of development consistent with the proposed models of redundant discrete systems.

§2. Basic Definitions, Concepts, and

Problem Formulations

In view of the wide diversity of terminology used by different authors we begin with the fundamental concepts and definitions to be used in the ensuing survey and analysis of the literature. The controlled object in an automatic system incorporating a discrete controller may be represented as a discrete model subject to all the basic tenets of redundancy theory with respect to functional reliability. In the present article, therefore, we use the term discrete system (see [142, 92, 143]) to connote the system complex formed by a discrete controller and controlled object without special designation of the position of the controller. This rationale is further justified by the fact that the present-day automatic control system usually contains a battery of controllers, which may be analyzed either collectively or individually. The most general model of a discrete system (finite automaton) is portrayed in Fig. 1.

We denote the set of inputs by $A = \{A_1, \ldots, A_m\}$, the set of logic elements by $X = \{X_1, \ldots, X_s\}$, the set of memory elements by $Y = \{Y_1, \ldots, Y_r\}$, and the set of outputs by $Z = \{Z_1, \ldots, Z_l\}$. If the

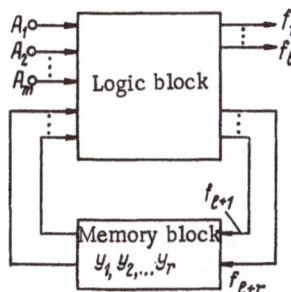

Fig. 1. Block diagram of
a discrete system.

device is a multicycle (sequential) machine, we have in general

$$z_i = z_i' (a, x, y); \quad i = 1, \ldots, l.$$

In the case of single-cycle (combinational) machines†

$$z_i = z_i'' (a, x).$$

It is important to note that the variables of the set x are not independent, but represent functions of the variables from a in the case of combinational devices or of the variables from a and y in the case of sequential devices, so that, depending on the type of device, either

$$z_i = z_i' (a, x(a, y), y)$$

or

$$z_i = z_i'' (a, x(a)).$$

In accordance with [142] the synthesis of a system means the determination of the system structure on the basis of its prescribed performance specifications.

We interpret the infallibility (or spatial reliability) of a discrete system as the property that the system executes the functional algorithm perfectly in the presence of simultaneous "failures" of a predetermined number of its elements or inputs [208]. We shall be concerned henceforth solely with discrete systems constructed from two-state elements, i.e., elements whose input and output functions can take only one of two possible values, which are conditionally assigned the numbers 0 and 1. In this case the failure of an element generates an incorrect realization of values at α outputs of that

---

†We denote the value of an input A_i and the states of elements X_i and Y_i or of an output Z_i by the corresponding lower-case letters a_i, x_i, y_i, z_i, and we denote the sets of the latter by $a = (a_1, \ldots, a_m)$; $x = (x_1, \ldots, x_s)$; $y = (y_1, \ldots, y_r)$; $z = (z_1, \ldots, z_l)$.

element ($1 < \alpha < \theta$, where θ is the number of outputs of the element);
specifically, a 1 may be realized in place of a 0 (in which case we
say that a $0 \rightarrow 1$ failure has occured at the given output) or a 0 may
be realized in place of a 1 (and we say that a $1 \rightarrow 0$ failure has
occurred at the given output).*

    If the probability of a $1 \rightarrow 0$ failure in a particular element
is of the same order as the probability of a $0 \rightarrow 1$ failure, we say
that the element is subject to symmetric failures. Otherwise it is
said to have failure assymetry. A distinction must be made between
failures of logic elements, memory (storage) elements, and inputs,†
on the one hand, and system output failures on the other. The introduction of redundancy (structural or information) can only prevent
failures in the first group. The influence of failures of the outputs
themselves (i.e., the circuits leading from the output element to
some nominal point of connection between the structure and external
circuitry) or of output elements (the terminal logic elements of the
output circuit) can be eliminated or reduced only in the next device
to which the given output is connected. The infallibility, or spatial
reliability, of a system can be characterized by one or two positive
integers. We say that a system is d-reliable [208] if for any d internal failures of the system‡ it still executes the functional algorithm perfectly but there exists a combination of d + 1 failures such
that the function algorithm is no longer correctly executed. We say
analogously that a system is (d_0, d_1)-reliable if for any $t_0 < d_0$ type
$1 \rightarrow 0$ internal failures and $t_1 < d_1$ type $0 \rightarrow 1$ internal failures the
system correctly executes the functional algorithm but there is at
least one combination of $d_0 + 1$ type $1 \rightarrow 0$ failures such that the
functional algorithm is not correctly executed for t_1 type $0 \rightarrow 1$
failures ($t_1 < d_1$) and there is at least one combination of $d_1 + 1$
type $0 \rightarrow 1$ failures such that the functional algorithm also fails for
t_0 type $1 \rightarrow 0$ failures ($t_0 < d_0$). The number d (or numbers d_0 and
d_1) and the probability of faultless operation for a definite period
of time are bound by a certain (specific to a given method) functional
dependence [93].

---

*Sometimes in the literature a $0 \rightarrow 1$ failure is called a "short-circuit" failure, and
  a $1 \rightarrow 0$ failure is called an "open-circuit" failure.
†In view of the fact that noise (errors) in the communication channel is a frequent
  source of failure at the inputs, input failures are also called errors.
‡We say that an internal failure occurs in the system if the failure is not a failure
  of the output proper or of some output element of the system.

One may speak in terms of the reliability of a system in relation to memory elements alone, assuming that the reliabilities of the system logic elements are so small as to be negligible. Similarly, one may speak of the reliability of a system in relation to logic elements alone or in relation to inputs alone.

The set of problems united by the common objective of investing a structure with a definite reliability may be formulated as follows.

1. To guarantee, in the synthesis of the discrete system, a prescribed functional reliability in relation to one of the sets A, Y, or X; in relation to the pairs of sets (A, Y), (A, X), or (X, Y); or in relation to all three sets of elements.† The reliability can be numerically specified either in terms of the number d of symmetric failures, in terms of the numbers d_0 and d_1 of $1 \to 0$ and $0 \to 1$ failures of elements, or in terms of the list of inadmissible failures. This same set of problems can also be stated in probabilistic terms.

2. To guarantee, in the synthesis of the discrete system, a prescribed probability of faultless operation for a definite period of time, given the known probabilities of faultless operation of the logic elements and memory elements in the same time period and known probabilities of errors in the sets of values of the input variables. In practice, however, the required probability of fault-less operation of the system often cannot be determined, whereupon the problem is to create a device with the minimum possible failure probability or maximum reliability subject to certain constraints on the weight or size of the system, its complexity, or the nature of the system structure (for example, construction of the system from identical blocks or modules that can easily be replaced in the event of malfunction, the incorporation of malfunction alarm systems, etc.). Consequently, the problem of assuring system reliability may be stated in the following alternative context.

3. To guarantee, in the synthesis of the discrete system, maximum reliability when specific constraints are imposed on the structure of the system. A distinction is made between stationary (or permanent) and transient failures. Transient failures differ from the stationary variety in that the distortion of the input or output value of a given element is of short duration (self-correct-

---

†In particular, a list of elements whose symmetric or asymmetric failures are inadmissible may be specified.

ing). Transient failures are also referred to in the literature as malfunctions. The only difference in the treatment of problems relating to the elimination of the influence of transient errors on system operation and the treatment of analogous problems for stationary failures is that only unit failure is deemed possible in the former, because the probability of the simultaneous occurrence of two or more transient failures is negligible. For this reason, in our survey of the literature we shall mention transient failures only insofar as they have bearing on methods that are specifically inapplicable to stationary failures.

One of the demands imposed on the reliable system is the absence of inadmissible races or hazards in its structure (i.e., system "stability" or "temporal" reliability). This condition is also a significant requirement in the synthesis of reliable discrete systems. Unfortunately, the simultaneous assurance of system reliability and stability has not received adequate attention heretofore. We shall examine the work done in this area in the third chapter.

§ 3. Redundancy Models

All extant methods for the introduction of structural redundancy may be grouped according to the model used to represent the redundant discrete system. We list the following categories of proposed models:

A. Models of redundant discrete devices using codes with replication:

1. Model with iterated structure (iteration model).
2. Model with majority structure (majority model).
3. Model with interwoven connections (interweaving model).

B. Models of redundant discrete devices using effective correcting codes (effective-coding models):

1. Model with error correction at the information inputs (input-correction model).
2. Model with the correction of memory element failures (memory-correction model).
3. Model with the correction of logic block failures (logical operation-correction model).
4. Model with mixed failure correction (hybrid correction model).

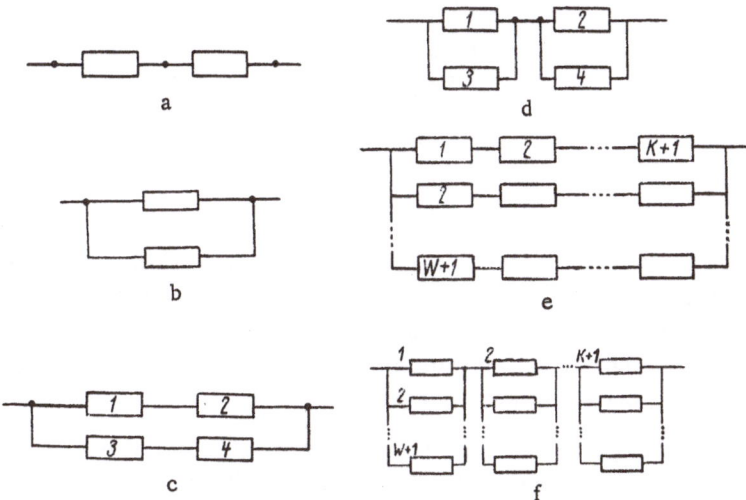

Fig. 2. Various types of series and parallel composition. a) 1-fold series composition; b) 1-fold parallel composition; c) 1 × 1 parallel-series composition; d) 1 × 1 series-parallel composition; e) w × k parallel-series composition; f) k × w series-parallel composition.

The iteration model was first proposed by Moore and Shannon [263]. The basic idea behind the construction of the iteration model is the replacement of an individual element or block forming part of the structure of a discrete system whose failures can be reduced to breakdown of the system functional algorithm by a set of several elements or blocks. The nature of the connection between elements of the set can differ depending on what types of failures are disallowed. Examples of such a model are found in the series† (Fig. 2a) or parallel‡ (Fig. 2b) composition of elements or blocks or cross-combinations thereof (see Figs. 2c, 2d, 2e, and 2f). The case of d_1-fold series composition affords $(0,d_1)$-reliability, and d_0-fold parallel composition affords $(d_0,0)$-reliability. The generalization of this composition method entails the replacement of elements (blocks) by a set of the same elements (blocks) with a definite type of connections between elements (blocks) of the set (some possible connections are shown in Fig. 3). This type of composition model was first investigated by Moore and Shannon [263]. They analyzed the behavior of the reliability function and proposed in the event of inadequate structural reliability the continued appli-

---

†Sometimes called "block" or "modular" composition in the literature.
‡Sometimes called "element" composition in the literature.

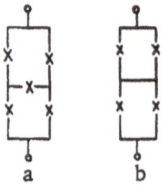

Fig. 3

cation of the same composition method, i.e., iteration. The number of iterations is arbitrary. It was also shown in [263] that the reliability function tends to unity as the number of iterations is increased without limit. The iterated composition model was subsequently investigated in several papers (see §4).

The m a j o r i t y m o d e l postulates the use of a special element or device called a majority organ. The operation of this element is described as follows. If the function ξ ($\xi \in \{0, 1\}$) operates on the "majority" of inputs of the element, the output must generate the function μ ($\mu \in \{0, 1\}$), where the significance of ξ and μ and the interpretation of "majority" are determined by the type of majority organ. In the most general case each input of the majority organ is assigned a definite weight, and its functions coincide with the functions of the so-called "threshold" element. In a special case the "majority" may be defined as $[k + 1]/2$, where k is the number of inputs to the majority organ. It is said in this case that the majority organ operates on the "vote-taking" principle. The majority model is applied to the system as a whole, to its most unreliable blocks, or directly to its elements. The part (block) of the structure to which majorization is applied is replaced by k similar blocks. The i-th output of every j-th block is connected to the j-th input of the i-th majority organ. Instead of the outputs of the original block the outputs of the majority organ are used. For the i-th output of each block the majority model can contain not one, but several majority organs. In this case the restoration of values by means of majority organs is realized simultaneously with multiplication of the outputs. Majorization can be applied in multiple succession.

Variants of the majority model are illustrated in Figs. 4 and 5.

The majority model was first proposed by von Neumann [269] and was later developed by several authors (see § 5). It is

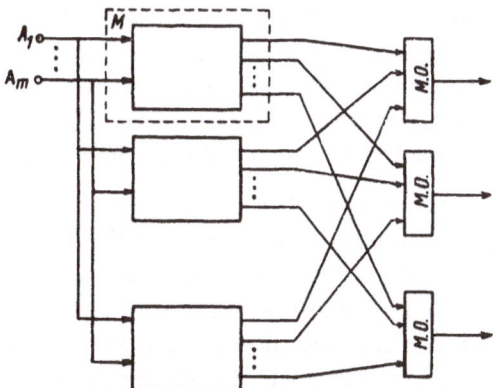

Fig. 4. (k,1)-structure majority model for a
multioutput system.

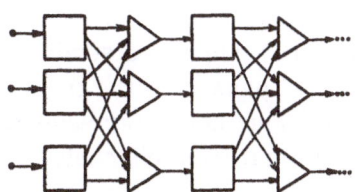

Fig. 5. Two-level (k, k)-structure
majority organ for a single-output
system.

important to note that the majority model is the most frequently
used in practical problems. Recently a majority model has been
developed with an adaptive restoring organ (see, e.g., Pierce's
book [278]). This topic will be discussed in closer detail in § 5.

The interweaving model was first proposed by
Tryon [300]. He described the structure of so-called modified
logic elements, which in series with a realization of the logic
function are capable of correcting certain errors at their inputs;
for example, in the case of an element realizing the function x =
$\overline{ab}$ the modified element will be one with tripled or quadrupled in-
puts realizing the function $x' = \overline{(a_1a_2 + a_1a_3 + a_2a_3)(b_1b_2 + b_2b_3 + b_1b_3)}$ or $x'' = \overline{(a_1a_2 + a_3a_4)(b_1b_2 + b_3b_4)}$.

Various techniques are possible for the construction of
the modified elements. The notion of "critical" and "subcritical"
failures† are introduced. A critical failure is a failure at the in-

---

†A subcritical failure is called a "subcritical error" in some papers.

puts of a given element such that an incorrect output value is realized. The minimum number of input failures generating an output failure is called the order of the critical failure. A subcritical failure is defined as a failure at one or more inputs of the element such that the element still generates the correct output value.

The logic block of a discrete device is constructed so that a critical failure in one layer (level) of the logical structure generates a subcritical failure in the next layer. The connections are constructed so that only some of the elements of the second layer will be subject to the action of a critical failure in the first layer. In the third layer subcritical failures occurring at the output of the second layer must be mixed with correct functions (signals) from the second layer, so that, as a result, failures are completely eliminated or diminished in number (the particular result depends on the mode of connection). The interweaving model is used on conjunction with modified elements.

The input-correction model appeared several years later with the publication of the first papers on the theory of effective† error-correcting codes and does not differ essentially from the use of correcting codes in communication channels.

The basic idea of the input-correction model is that the initial sets of values of the input variables a_1, ..., a_m are assigned an n-dimensional correcting code (n > m). The coding operation produces additional sets of symbols a_{m+1}, ..., a_n, which are transmitted to additional inputs A_{m+1}, ..., A_n. The reproduction of the values of the output functions is realized by means of additional checking and correcting blocks. The latter can exist independently or in composition with the initial blocks. Inasmuch as the input values can be distorted by noise in the communication channel by which they are transmitted, the encoder is inserted before the communication channel and after the message source or in coexistence with the latter (see Fig. 6).

One of the earliest investigations of the input-correction model was Peterson and Rabin's paper [273].

The memory-correction model was first proposed by Gavrilov [9]. It works as follows. The set of element states

---

†We use the term "effective" throughout this survey to mean minimum-redundancy error-correcting codes.

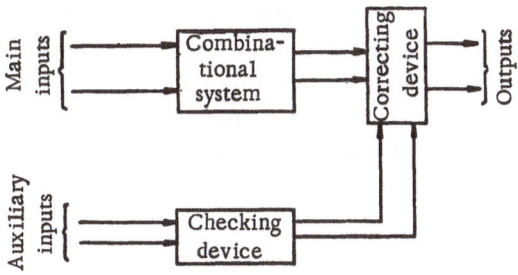

Fig. 6. Input-correction model.

described by the vectors $(y_1, ..., y_r)$ is assigned a linear correcting code, producing a set of n-dimensional vectors $(y_1, ..., y_r, y_{r+1}, ..., y_n)$. On account of the linearity of the code each auxiliary component y_{r+i} represents a linear combination of the primaries. The relations between the functions y_{r+i} and the functions y_j ($j = 1, ..., r$) establish Boolean expressions for all components of the n-dimensional vector, i.e., for all functions realized by the feedback relations. Consequently, the encoder coincides with the logic block. Since the decoder also coincides with the logic block, the Boolean expressions realized by the outputs are varied with regard for the particular correcting code used. The block diagram of the model differs from the one shown in Fig. 1 only in the number of feedback links.

Following the publication of Gavrilov's investigations the memory-correction model was studied by several authors. It began to be used for the correction of failures of various types of elements used in memory devices and to be combined with other types of models of redundant discrete devices. This problem will be discussed in detail in § 7.

The logic-correction model was first investigated by Elias [196], and its specific realization for the creation of a d-reliable contact structure, with d = 1, realizing one or more Boolean functions was investigated by Zakrevskii [35]. A block diagram of the model is given in Fig. 7. In the first-mentioned paper the effectiveness of coding for the enhancement of the reliability of a logic converter was investigated. The principle of the model is as follows. An unreliable combinational system is treated as a perfectly reliable system executing a given functional algorithm coupled in series with a noisy communication channel by which the system output states are transmitted. Since the reliability of transmission of discrete messages is improved by the use of a correcting code, the sets of values $(z_1, \ldots, z_l)$ is encoded by a linear correcting code. The coding operation matches the l-dimensional vectors $(z_1, \ldots, z_l)$ with n-dimensional vectors $(z_1, \ldots, z_l, z_{l+1}, \ldots, z_n)$, where

$$z_j = \sum_{i=1}^{l} \oplus \alpha_i z_l \quad (j = l+1, \ldots, n).$$

The additional functions z_j $(j = l + 1, \ldots, n)$, following substitution therein of the values of the primary functions, are transformed into functions dependent on the same input variables as the primary functions. In order to exclude dependent distortions at the inputs all functions are realized independently by single–input blocks, increasing the number of inputs. The set of these blocks is simply a device incorporating the initial logic converter and an encoder. A correcting block of the same kind as those used in communication systems with correcting codes is used for reproduction of the values.

In subsequent papers this model was further developed and generalized to meet various encoding techniques. This problem will be considered in detail in § 7.

The hybrid correction model represents a combination of two or three different models and is therefore described by any of several block diagrams. One example is the block diagram illustrated in Fig. 8, which shows a combination of three models: memory correction, logic correction, and input correction (see [57]).

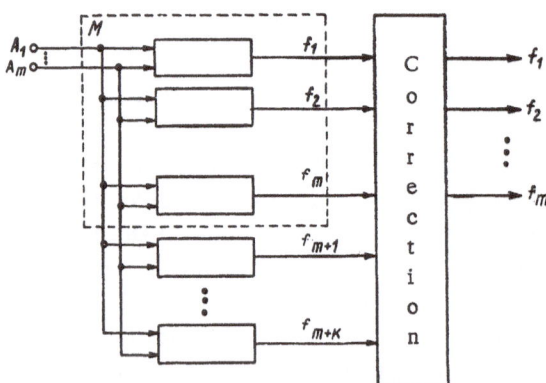

Fig. 7. Logic-correction model.

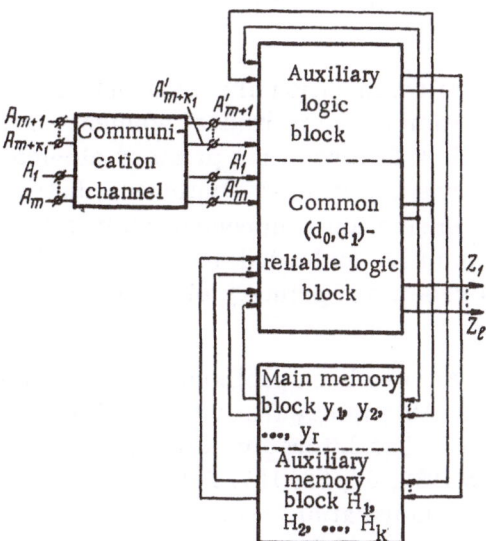

Fig. 8. Simultaneous memory-, logic-,
and input-correction model.

§4. Composition Methods

As we mentioned above, reliability-improvement methods based on active or passive composition (replication) have long been in existence. However, the theoretical aspects of this approach have been under serious investigation only in about the last forty years, on account of the advent of various types of power and control systems and, accordingly, the higher demands for reliability assessments. According to the Soviet State Standard (GOST) definition of composition [140] this method covers any reliability-improvement methods involving the introduction of redundancy. In the present survey, however, we shall interpret composition in the sense of specific modes of redundancy, namely the series addition of elements (blocks, subsystems) for $0 \to 1$ failures, the parallel addition of elements (blocks, subsystems) for $1 \to 0$ failures, and the parallel-series addition† of elements (blocks, subsystems) for both types of failures in combination, as well as composition by the iteration model. Composition can be effected either with respect to some or all elements of the system, or with respect to the sys-

---

†We use the notion of parallel-series composition categorically to include series-parallel composition as well.

tem as a whole. Depending on which it is, we call it elemental or general composition.

Two types of composition are discerned according to the time at which operation of the redundant elements is initiated (concurrently with or after failure of the primary elements): we define composition as (a) passive (loaded, continuous, hot) when the redundant or reserve elements function concurrently with the primary elements, or as (b) active (unloaded, cold)[†] when the redundant or reserve elements supersede primary elements in the event of failure of the latter.

Many papers have been published on both types of composition, and it is reasonable to say that these problems have been adequately investigated. The Literature Cited lists only the most important references [183, 87, 33, 107, 20] in which the reader can find comprehensive information. In these references expressions are investigated for determining the reliable-operation characteristics of discrete systems, such as the probability of faultless operation, the mean time of reliable operation, etc. In view of the fact that simple (series or parallel) composition in complex discrete systems has been replaced in recent years by more precise methods we shall describe only the basic equations for the probability of faultless operation for those methods, focusing our attention primarily on investigations that deal with the more complex redundancy methods.

Series composition means that the element or block for which it is desired to minimize the effect of failures are connect ed in series with a chain of k identical elements or blocks in succes sion. This system will correctly realize an output of 0 as long as at least one of the k + 1 elements or blocks does not fail. The order of the composition in this case is equal to k. A block diagram of the series composition mode for k = 1 is shown in Fig. 2a. We denote the probability of faultless operation of the i-th element (block) by $p_i(t)$, whereupon the probability $P(t)$ of faultless operation of the total system is expressed by the equation

$$P(t) = \prod_{i=1}^{k+1} p_i(t),$$

[†]This type of composition is sometimes called substitutional composition.

and the probability of failure is given by the equation

$$Q(t) = 1 - P(t) = 1 - \prod_{i=1}^{k+1}(1 - q_i(t)).$$

In the special case in which

$$p_1(t) = p_2(t) = \cdots = p_{k+1}(t) = p(t) = 1 - q(t), \tag{1}$$

we obtain

$$P(t) = \{p(t)\}^{k+1} \text{ and } Q(t) = 1 - (1 - q(t))^{k+1}.$$

Parallel composition entails the connection of w identical elements (blocks) in parallel with the primary element (block). The order of the composition in this case is equal to w. The faultless-operation and failure probabilities in this case are expressed by the equations

$$P(t) = 1 - \prod_{i=1}^{w+1}(1 - p_i(t)); \quad Q(t) = 1 - P(t) = \prod_{i=1}^{w+1}q_i(t)$$

or in the case of (1)

$$P(t) = 1 - (1 - p(t))^{w+1}, \quad Q(t) = \{q(t)\}^{w+1}.$$

Parallel-series composition represents the combination of the two types of composition described above. Two variants of parallel-series composition are illustrated in Figs. 2e and 2f. The expression for the faultless-operation probability depends on the sequence in which composition is realized. For the variant of Fig. 2e

$$P(t) = 1 - \left[1 - \prod_{i=1}^{k+1}(1 - q_i(t)) \right]^{w+1}$$

or in the case of (1)

$$P(t) = 1 - |1 - (p(t))^{k+1}|^{w+1}.$$

For the variant of Fig. 2f

$$P(t) = \left\{ 1 - \prod_{i=1}^{w+1}[1 - p_i(t)] \right\}^{k+1}$$

or in the case of (1)

$$P(t) = |1 - (q(t)^{w+1}|^{k+1}.$$

In the book by Gnedenko, Belyaev, and Solov'ev [20] those variants are compared in terms of the faultless-operation time.

We assume for simplicity that k = w = 1 (see Figs. 2c and 2d).
We denote the faultless-operation time of the i-th element by τ_i,
the faultless-operation time of the total system based on Fig. 2c
by $T_{(c)}$, and the faultless-operation time of the total system based
on Fig. 2d by $T_{(d)}$, whereupon

$$T_{(c)} = \max[\min(\tau_1, \tau_2), \min(\tau_3, \tau_4)],$$

$$T_{(d)} = \min[\max(\tau_1, \tau_3), \max(\tau_2, \tau_4)],$$

but $\min(\tau_1, \tau_2) \leq \tau_1$ and $\min(\tau_1, \tau_2) \leq \tau_2$, while, analogously,
$\min(\tau_3, \tau_4) \leq \tau_3$ and $\min(\tau_3, \tau_4) \leq \tau_4$, so that

$$T_{(c)} < \max(\tau_1, \tau_3), \quad T_{(c)} < \max(\tau_2, \tau_4),$$

hence

$$T_{(c)} < \min[\max(\tau_1, \tau_3), \max(\tau_2, \tau_4)] = T_{(d)}.$$

Consequently, composition by the circuit of Fig. 2d is
more favorable. However, this result is not always valid, as indicated by Polovko [107], if the elemental realization of both structures is taken into account.

Parallel-series composition of order k = w = m has been
investigated by Potapov and Yablonskii [109] for the solution of
the problem of assuring the m-reliability of a universal structure
capable of realizing any Boolean function of n variables. In this
case the complexity $L^m(n)$ of the structure satisfies the relation

$$L(n) < L^m(n) < (m+1)^2 L(n).$$

If, on the other hand, this type of structure, which is in
fact (m,m)-reliable, is used to correct only $0 \to 1$ failures, its
complexity $L_3^m(n)$ can be reduced:

$$L_3^m \leq \left\{\left[\frac{m}{2}\right] + 1\right\}\frac{2^n}{n}$$

[since the complexity L(n) of the realization of any Boolean function
of n variables is $L(n) \sim 2^n/n$].

Madatyan [75] has investigated the complexity of the realization of a $(d_0, 0)$-reliable single-output contact structure realizing
any of 2^{2^n} Boolean functions of n variables. The structure is synthesized by the same methods as that used to determine the complexity of the realization of nonredundant structures. It is shown

that the complexity $L_p^1(n)$ of the indicated structure, expressed in the number of contacts, satisfies the following relation for $d_0 = 1$:

$$L_p^1(n) \leqslant \frac{2^n}{n},$$

whereas in the case $d_0 > 1$ the complexity $L_p^{d_0}(n)$ is estimated by the expression

$$L_p^{d_0}(n) \leqslant \frac{d_0+1}{2} \cdot \frac{2^n}{n}.$$

Nechiporuk [88] has studied the realization complexity of two-layer diode structures that realize (p,q)-matrices and are $(d_0,0)$-reliable (with respect to the diodes). It is shown that, given unbounded growth of p and q, the complexity of the structures for "almost all" matrices is at least c(m + 1) times the complexity of conventional structures $\left(c = \frac{1}{e \ln 2} \right)$. If the matrices contain only a few ones, the given structures are asymptotically m + 1 times as complex as conventional structures for almost all matrices.

The parameters d_0 and d_1 of (d_0,d_1)-reliable contact, diode, and diode-contact structures for which the asymptotic behavior of the Shannon functions is nonincreasing have been investigated in two other papers by the same author [89, 90]. In the first paper two methods are considered for the synthesis of (d_0,d_1)-reliable diode and contact structures without the use of correcting codes, where the asymptotic value of the Shannon function is nonincreasing or at most doubles. In the second paper [90] it is shown that the asymptotic value of the Shannon function is nonincreasing for all $(d_0,0)$-reliable contact structures with $d_0 = o\left(\frac{\log n}{\log \log n} \right)$ and for almost all $(d_0,0)$-reliable contact structures with $d_0 = o\left(\left(\frac{n}{\log n} \right)^{1/2} \right)$. The asymptotic value of the Shannon function doubles at most for (d_0,d_1)-reliable contact structures $d_0 = o\left(\frac{\log n}{\log \log n} \right)$ and $d_1 = \left[n^{\frac{1}{2}-\varepsilon} \right]$, where ε is an arbitrarily small fixed number.

For (d_0,d_1)-reliable parallel-series contact structures with $d_0 = o\left(\left(\frac{\log n}{\log \log n} \right)^{\frac{1}{2}} \right)$ and $d_1 = o\left(\left(\frac{\log n}{\log \log n} \right)^{\frac{1}{2}} \right)$ and for diode-contact structures with $d_0 = o\left(\left(\frac{n}{\log n} \right)^{\frac{1}{2}} \right)$ and $d_1 = o\left(\left(\frac{n}{d_0 \log d_0 + \log n} \right)^{\frac{1}{2}} \right)$ the asymptotic behavior of the Shannon function is nonincreasing. The method of synthesis of (d_0, d_1)-reliable diode structures is based on the multiple covering of the Boolean matrices by their unit submatrices. A necessary condition is found in order for the covering chain to be nonincreasing with increasing multiplicity.

It is appropriate to include here the work of Rabinovich [113] on the problem of synthesizing minimal $(d_0, 0)$-reliable contact structures with $d_0 = 1$ realizing Boolean functions of three variables.

In [112] Rabinovich determines estimates of the realization complexity of a Boolean function f of n variables, $f(x_1, ..., x_n) = x_1 \oplus x_2 \oplus ... \oplus x_n$. It is established that the realization complexity $L_{p3}^{11}(n)$ of (1, 1)-reliable contact structures is 8n.

Malyugin discusses in [78] a technique for increasing the faultless operation probability by a special mode of composition taking into account the specified list of failures. Instead of the given function a structure is realized according to the scheme of Fig. 2c, in which block 1 realizes F_1 and the other blocks realize certain other functions F_2, F_3, and F_4. Each of the latter ensures the correct realization of the functional algorithm for different failures included in the list. It is shown that the resulting structure is less complex than under the realization $F_1 = F_2 = F_3 = F_4$.

Lowrie [247] has investigated a technique for increasing the reliability of computers on the basis of duplication with cross-connection (duplex composition) in combination with a diagnostic program for determining the failure location.

In a paper by Fan Liang Tseng, Wang Chiu Sen, F. A. Tillman, and Hwang Ching Lai [198] the order of composition is determined on the basis of a cost comparison of failure-induced down times and the incorporation of redundancy. Investigations in this same general area are described by Babaev and Babaeva [3].

Mine and Koga [261] have sought to ascertain the properties that a realizable Boolean function must have in order for its realization to be d-reliable for a finite redundancy.

Fractional-order composition refers to the composition technique in which the redundant system consists of n separate systems and requires for normal operation the correction of at least k systems. The order of the composition in this case is equal to (n − k)/k. This type of composition is effective for systems subject to different kinds of failures, particularly in the case of high-reliability systems designed for short-term operation. The results of a study of fractional-order composition are described in a paper by Polovko [107].

Iterative composition is a method that was proposed by Moore and Shannon [263] for structures synthesized from contacts (two-way conducting elements). The method was subsequently generalized by several authors to structures synthesized from certain types of contactless elements. The method proposed in [263] calls for the replacement of a contact subject to failures (permanent or transient) by a certain structure (examples of which are shown in Figs. 3a and 3b) synthesized from contacts of the same relay or its repeaters.† Each of these structures correctly realizes an output value of 0 or 1 for any failure of any type and for certain pairs of failures. The same procedure is applied to each contact of the resulting structure in order to enhance the degree of reliability, i.e., iteration is effected. The behavior of these structures was investigated by analyzing the function h(p) representing the probability of the structure being in the conducting state (closure function) as a function of the probability p of any contact being in the conducting state. The following assumptions were adopted: (1) the failures of a particular contact in different time intervals are mutually independent; (2) the failures of different contacts are mutually independent; and (3) the probabilities of $1 \rightarrow 0$ and $0 \rightarrow 1$ failures are equal for all relay elements, independent of the choice of structure, and invariant with time.

Inasmuch as a contact is not absolutely reliable, when the relay element is in the unexcited state the contact is closed with a probability a and open with a probability $1 - a$; when the relay element is in the excited state the contact is closed with a probability c and open with a probability $1 - c$. If $a < c$, the contact is closing, and if $a > c$, the contact is closing, and if $a > c$, it is opening. Contacts for which $a \neq 1$ and $c \neq 0$ (or $a \neq 0$ and $c \neq 1$) are said to be unreliable.

---

†The relay windings are assumed to be absolutely reliable.

In the general case

$$h\,(p) = \sum_{i=0}^{n} H_i p^i\,(1-p)^{n-i},$$

where n is the total number of contacts in the structure and H_i is the number of ways in which a set of i contacts can be chosen from a total set of n in order for the structure to be in the conducting state when the i contacts chosen are closed and the remaining n - i are open. The opening probability function is written analogously:

$$1 - h\,(p) = \sum_{i=0}^{n} B_i\,(1-p)^i\,p^{n-i}.$$

The function h(p) has the following values for the indicated structures (Figs. 2a and 2b): $h(p) = 4P^3 + p^4$ for structures (a) and $h(p) = 2p^2 + 2p^3 - 5p^4 + 2p^5$ for structure (b). Other structures can be used besides those in Figs. 2a and 2b.

We have the following for the resultant structure of two structures with functions $h_1(p)$ and $h_2(p)$ in series or parallel:

$$h(p) = h_1(p) \cdot h_2(p) \quad \text{and} \quad h(p) = 1 - (1 - h_1(p))(1 - h_2(p)).$$

In addition to the parallel or series connection of two structures we can have an "iterative" connection, in which each contact in the first structure [for which the probability of the conducting state is equal to $h_1(p)$] is replaced by the second structure [with conducting-state probability function $h_2(p)$]. The resultant function h(p) than has the form

$$h(p) = h_1(h_2(p)).$$

If this process of iteration repeats n-1 times, then the nth iteration of the function h (p) has the form

$$h_{(n)}(p) = h(h(\ldots(h(p)))).$$

As shown in [263], $h_{(n)}(p)$ tends to unity as the number of iterations is increased without limit. Also investigated in the same paper is the increase in the realization complexity due to iteration; the following is proved in particular:

Theorem. If the number $\delta > 0$, $0 < a < c < 0$, and $d = \max\left(\frac{a+c}{2},\ 1 - \frac{a+c}{2}\right)$ are given, there is a structure for which $h(a) < \delta$, $h(c) > 1 - \delta$, containing at most N contacts, where

$$N = 81 \left[\frac{\log \frac{c-a}{4}}{\log d} \right] \left(\frac{1}{c-a} \right)^{\frac{\log 9}{\log \left(\frac{3}{2} \right)}} \left(\frac{\log \sqrt{8\delta}}{\log \sqrt{8}} \right)^2 .$$

A series of papers were published in the wake of Moore and Shannon's paper [263], continuing or further generalizing the study of parallel-series and iterative composition. Esary and Porschan [197] showed how, using the concepts of minimal paths and minimal cuts, which are correlated by Boolean functions, to find upper estimates of the probability of faultless operation of a structure. This approach makes it possible to compute a function analogous to h(p) but more general in that the structure can contain contacts of different relays having different closure probabilities.

Barlow, Hunter, and Proschan [171] proposed a technique for determining a structure of the type shown in Fig. 2a with an optimum number w + 1 of parallel branches for a fixed number k +1 of contacts of each branch. They also showed that the failure probability of the structure in Fig. 2e can be made as small as one wishes by a suitable choice of w and k, provided the failure probabilities of the relays themselves are sufficiently small.

Pierce [278] has given an alternate proof of the fact that the structure of Fig. 2e can be made as reliable as one wishes; he gives the following relation between w and k:

$$k + 1 = \frac{-2 \log (w+1)}{\log (q_1 p_2)} . \tag{2}$$

Here $q_1 = 1 - p_1$, p_2 is the probability that the relay contacts will be closed when they are supposed to be open, and p_1 is the probability that the relay contacts will be open when they are supposed to be closed. Under relation (2) it is guaranteed that the failure probabilities of this structure will vanish as k, w → ∞.

In [278] Pierce also gives an estimate of the failure probability Q of any relay structure containing N contacts, whereby

$$\ln Q \sim \sqrt{N} \ln c,$$

where c is a constant.

The technique of Moore and Shannon has been used in a number of papers for other types of elements or has been elaborated

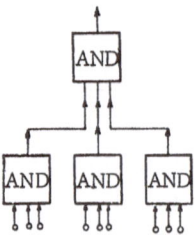

Fig. 9. A typical
recursive triangular
switching network.

under more general assumptions. Wallmark and Revest [307], for
example, generalize the Moore—Shannon model to transistors,
analyzing their reliability. Hall [215] and Domanitskii and Prangish-
vili [29] have conducted investigations in the same vein. In the former paper transistorized inverter, trigger, and monostable multi-
vibrator circuits are proposed which are insensitive to any single
failure or certain pairs of failures; in the latter paper a circuit is
developed for a diode-transistor NOR element insensitive to any
single failure or certain pairs of failures of its components.

Another, somewhat different iteration model has been developed by Levy [240]; specifically, he uses recursive triangular
switching networks, an example of which is illustrated in Fig. 9.
These circuits have also been used by Amarel and Brzozowski [165].

In [152] Fabrikant attempts to apply the Moore—Shannon
model to the case of unreliable relay windings. Kochen [238] has
determined the dependence of the number of redundant relays in the
Moore—Shannon model on the logic function of a particular network
and analyzed the reliability characteristics of elementary AND, OR,
and exclusive-OR structures, the characteristics of the number of
relays, the topology of the network, and the distribution of inputs.
He also proposes a general method for determining the faultless-
operation probability of a combinational switching network.

Blake [177] has proposed a method for determining the
failure probability of a hammock network with regard for the types
of failures. The reliability function defined by Moore and Shannon
has been studied by Reza [284]; he expands it in the functions
$f_{\tau_1, \ldots, \tau_k}$, where $f_{\tau_1, \ldots, \tau_k}$ is the reliability function of the same
structure when the branch τ_j ($j = 1, \ldots, k$) is replaced by a busbar
or is deleted.

Two other papers in which the results of Moore and Shannon are developed should be mentioned: one by Urbano [304] and another by Petri [101]. In the former the primary structure is used as the iterative element. The theoretical foundations of this type of iteration is given. It is shown in the latter paper that the minimum number $L(\rho, n, \varepsilon)$ of contacts in a structure synthesized by the Moore — Shannon method realizing any function of n variables satisfies the relation

$$L(\rho, n, \varepsilon) \sim n \frac{\log^2 \varepsilon}{\log^2 \rho},$$

provided $\rho \leq 1/4$, $n \geq 2$, and $\varepsilon \leq \varepsilon_0(\rho, n)$. Here ρ is the failure probability of the contact, ε if the failure probability of the iterated contact, and $\varepsilon_0(\rho, n)$ is a positive definite function with values close to zero.

§5. Majority Methods

The original source for the formulation and investigation of majority methods was von Neumann's paper [269]. The redundancy scheme proposed by von Neumann is illustrated in Fig. 4. In this method an initial system with l outputs is replaced by n (n = 3 or, in general, n > 3 †similar systems controlled by the same output functions. The i-th output of the j-th system (for all i = 1, ..., l and j = 1, ..., n) is connected to the j-th input of the i-th of special organs called mixers.‡ The i-th mixer realizes the so-called "majority function," i.e.,

$$f = b_1 b_2 + b_1 b_3 + b_2 b_3,$$

where the inputs of the majority organ are designated b_1, b_2, and b_3, respectively.

The outputs of the majority organs are then treated as the outputs of the system. This method was proposed for systems constructed from unilateral conducting elements. Estimates of the faultless-operation probability of the system are also given in von Neumann's paper.

---

†The method is customarily referred to as triplication in the case n = 3.
‡In subsequent developments of the von Neumann method this organ is called a " vote-taking" or " majority" organ.

Let η_i be the probability that the input value b_i of the majority organ is false, so that the upper limit θ of the probability that at least two inputs have false values is

$$\theta = \eta_1\eta_2 + \eta_1\eta_3 + \eta_2\eta_3 - 2\eta_1\eta_2\eta_3$$

or, for $\eta_i \leq \eta$,

$$\theta \leqslant 3\eta^2 - 2\eta^3.$$

Let ε be the probability of a false output by the majority organ when its input values are correct. Then the failure probability at one output of the majority redundant system is equal to

$$\eta^* = [\varepsilon(1-\theta) + (1-\varepsilon)\theta](3\eta^2 - 2\eta^3) + \varepsilon = \varepsilon + (1-2\varepsilon)(3\eta^2 - 2\eta^3).$$

Analysis of the function $\eta*$ revealed that the failure probability can be diminished by iteration of the method described above only if the requirement $\varepsilon < 1/6$ is fulfilled, i.e., if the unreliability of the majority organ is bounded. A multiple transmission line method was also proposed by von Neumann, based on the notion that each communication line by which effects are transmitted between elements, blocks, or subsystems is replaced by a bundle of N lines. A certain critical (confidence) level Δ ($0 < \Delta < 1/2$) is selected, and the bundle is assumed to be in the excited state if the number of excited lines is $\geq (1 - \Delta)N$ and in the unexcited state if $\leq \Delta N$ lines are excited. Excitation levels between these two levels are regarded as intermediate or uncertain. Two types of elements must be present in the network, viz., the actuating organs, which exercise the main control operation over the bundles, and the restoring organs, which restore the excitation level of the bundles. The functions of the restoring organ can be executed by any mixer as long as an N-fold branching of its output lines is effected.† It is assumed that the excited or unexcited lines are randomly distributed in the bundle. The majority organ can be synthesized from the same elements as the main system. Von Neumann [269] showed that the use of sufficiently many bundles of lines makes it possible by the given method to reduce the probability of system output error to as small a value as desired (provided ε is small enough). It is also required to make an appropriate choice of the confidence level Δ. However, von Neumann himself ques-

---

†We shall therefore also refer to the restoring organ hereinafter as a majority organ.

tioned the practicability of the method, because it increases the
number of lines by a factor of N and increases the number of primary organs by a factor of 3N, and in the examples that he investigated N came to about 20,000. The principles and notions discussed in the paper provided the impetus for a whole new field of
research. Many papers were published in which von Neumann's
methods were applied to specific types of elements and investigated from the point of view of admissible reliability, etc. The papers
representing direct continuations of von Neumann's work may be
grouped into an independent category, including papers by Verbeek
[305], Nadler [268], Teoste [296], Liu Chung [244], Muroga [267],
and Svechinskii [137, 138].

In [305] Verbeek investigates both methods of von Neumann
in application to structures composed of threshold elements. The
majority organs are also synthesized from threshold elements and
have m inputs. It is shown that the faultless–operation probability
of the structure is greater than the corresponding criterion for the
initial structure if redundancy is introduced by the scheme of Fig.
5 and one majority organ is used at the structure input. The failure
probability ε of the threshold element in this case, i.e., the probability that its threshold θ will assume either of the two values $+\infty$
or $-\infty$, is

$$\varepsilon < \frac{1}{2} - \frac{2^{m-2}}{mC_{m-1}^{\frac{m-2}{2}}}.$$

An analogous scheme is used by Nadler [268] for the intro-
ducntion of redundancy. In addition, a failure signalizer is used.
The probability characteristics of faultless operation are estimated
in connection with the proposed redundancy technique. A majority
model using majority organs synthesized from threshold elements
has been investigated by Lowenschuss [246]. Liu Chung [244] and
Muroga [267] also apply the majority model to threshold-element
structures. It is shown by the latter author that the redundancy
introduced into the structure by the use of threshold elements as
the majority organs is less than in von Neumann's circuits. In the
other two papers cited [137, 138] Svechinskii uses the multiple-line
method to synthesize multicycle relay devices that are insensitive
to malfunctions of d elements internal to the structure.

Another extension of the majority model has been given by
Mullin [266]. He uses stochastic logic, which makes it possible to

consolidate the classical logic-probabilistic theories of von Neumann
and McCulloch [255], on the one hand, and the classical probability-
theoretic calculus of Moore and Shannon on the other, with subsequent development in several models. Work has also been done
along this line by Tsertsvadze [157], who treats the failure-prone
automaton as a stochastic system.

One group of papers is concerned with the study of the conditions that must be satisfied by the failure probabilities of majority
organs and elements of the primary structure in order for a desired
faultless-operation probability of the redundant structure to be
achieved. The most rational network topology has also been investigated with a view toward maximizing the faultless-operation probability for a fixed redundancy or minimizing the latter for a fixed
faultless-operation probability. This particular group includes the
work of Lyons and Vanderkulk [248], in which the triplication method
is applied to a structure as a whole as well as to its individual modules (triple-modular redundancy). The conditions are determined
that must be satisfied by the failure probability in order for triplication of the structure as a whole not to require a greater redundancy than triplication applied to the individual modules. Kemp
[233] has investigated the probabilistic reliability characteristics
of a majority organ and determined their sufficient limit in order
to obtain a given faultless-operation probability for the structure.
In [190] Deo determines the failure probability of a majority structure with an absolutely reliable majority organ as a function of the
number of "vote-takers" (i.e., the number of redundant systems)
and shows that if the total number of vote-takers is 2k + 1, the
failure probability of the global system comprising 2k + 1 primary
systems is

$$Q_{2k+1} = \frac{(2k+1)!}{k!} q^{k+1} \sum_{\nu} \frac{(-q)^{\nu}}{\nu! \, (k-\nu)! \, (k+\nu+1)!},$$

where q is the failure probability of the primary system and

$$Q_{2k-1} - Q_{2k+1} = C_{2k-1}^{k} q^{k} (1-q)^{k} (1-2q).$$

This expression can be used to determine the best number k to
which the number of primary systems should be increased in the
majority model (without regard for the reliability of the majority
organ). Similar investigations have been carried out with regard

for the failure probability of the majority organ by Gol'denberg and Malev [21].

Mann [250] has explicated the principal factors governing the choice of network for the introduction of redundancy by the majority model. The most significant factor is the faultless-operation probability of the majority organ. The larger its value, the greater will be the number of majority organs in the structure of the redundant system in order to increase its faultless-operation probability, as for example in the application of the majority model to each element of the primary system. Equations are derived relating the faultless-operation probability of the structure to the faultless-operation probability of the majority organ when the latter is also constructed using the majority model. These investigations are continued in another paper by the same author [251], in which, specifically, majority organs having different structures are studied. One of these is the "transor" structure developed by Mann, which utilizes the ideas of potential-pulse elements, i.e., elements responding to a change of state of the inputs. For every time interval of fixed length the "transor" computes the number E_t of binary units at its inputs and compares them with the analogous number E_{t-1} computed for the preceding time interval. If the change $E_t - E_{t-1}$ is positive and greater than an assigned threshold T, the "transor" generates a unit at the output. If, on the other hand, the change is negative, $E_t - E_{t-1} < 0$, and its absolute value is greater than T, a zero is generated at the "transor" output. In all other cases the output value is left unchanged.

Also included in this group of papers is the work of Domanitskii [27]. He investigates the possibility of obtaining a failure probability arbitrarily close to zero for various types of structures at the expense of an increase in redundancy for a fixed reliability of the logic elements of which the redundant structures are composed. It is shown that for both conventional and adaptive redundant structures there exists a limiting faultless-operation probability, $P_0 < 1$, such that any variation of the redundancy cannot produce a structure having a faultless-operation probability $P > P_0$. The limit P_0 is determined by the faultless-operation probability of the majority organ. Special majority organ structures synthesized from NOR elements are investigated in this connection.

A separate group is made up of papers in which prior to application of the majority model the system is partitioned into blocks, and the majority model is applied to these blocks. Farrell [199] has investigated the structure of the kind of majority organ best suited (in terms of reliability criteria) to this application, analyzed the distribution of redundancy among the blocks, and sought the most effective block partition. Similar investigations, but for systems synthesized from cryogenic elements, have been reported by Griesmer, Miller, Roth, and Thomas [214]. Checking devices are used to tie together the blocks to which the majority model is to be applied. It is shown that the faultless-operation probability of systems of cryogenic elements can be increased perceptibility for a high enough faultless-operation probability of the elements ($P \sim 10^{-5}$). The optimum block partitioning of a system has also been sought by Domanitskii [27] for two types of majority models (with one or more majority organs for each output) using both conventional and adaptive structures.

In [235] Klaschka uses only "partial" redundancy, which connotes triplication of the part of the system having the greatest influence on the faultless-operation probability of the total system. The plane (q, k) is analyzed, where q is interpreted as the failure probability of the initial system, $k = q_2/q_1$, and q_1 and q_2 are the failure probabilities of the triplicated and untriplicated parts of the system, respectively. The following results are obtained: 1) In the plane (q, k) the domain is found in which partial redundancy affords a higher reliability than total triplication; 2) the value of k is determined which minimizes the failure probability of the system for fixed q; 3) the minimum value of the failure probability using partial redundancy is determined in the domain in which it affords a higher reliability than total triplication.

Block partitioning of a system has also been proposed by Lysikov and Mamedli [73], but different redundancy models are applied to different blocks. The system investigated in [73] is a computer represented in the form of two blocks, as shown in Fig. 1. It is proposed that the reliability of the logic block be increased by composition, and that of the memory block by majorization. Both redundancy techniques are realized at the functional block level.

In [283] Repton describes a program for determining the most effective position of a majority organ in order to maximize system reliability. Similar investigations have been conducted by dePian and Grisamore [191, 274] as well as Diduk [26], who applies

the majority model at the element level and uses threshold elements
for the majority organs. In this method, however, the structural
redundancy is very high, and the author himself expresses the opin-
ion that the method can only be used for microminiaturized electronic circuits.

In view of the significance of the faultless-operation probability of the majority organ in the reliable operation of a majority
structure many investigations have been made with an attempt to
increase the reliability of the majority organ or to determine relations by which it would be possible to exhibit the efficacy of a majority organ with fixed reliability characteristics. Ord-Smith, [270],
for example, describes a special type of majority organ. Let t be
the number of errors that can occur at the inputs of a k-input major-
ity organ, $0 \le t \le k$. In the conventional majority organ, if $t < \frac{1}{2}(k + 1)$
the failures are corrected, otherwise they are not. The majority
organ whose circuit is described in [270] has the property that fail-
ures can still be corrected for $t \ge \frac{1}{2}(k + 1)$ if $\lambda < \frac{1}{2}(k + 1)$, where
λ is the number of branchings of each output of the initial system
when the majority model is applied to it. In other words, the initial
single-output system is replaced by μ replicas, which are not connected to it and whose outputs have a branching factor λ. A generalized majority organ represents the consolidation of c conventional majority organs, each of which has b outputs, so that k =
$\lambda \mu$ = bc, but the numbers of branchings λ is always strictly less
than c.

The properties of majority organs are investigated in the
previously cited study of Lowenschuss [246]. It is shown that threshold elements sometimes work well as majority organs.

Also included in this group of papers are the investigations
of Gribanov [22], Braslavskii [7], Litvinov [70], and Savchenko [127].
In [22] special types of majority organs are analyzed which are
synthesized from unilateral and bilateral conducting elements and
in which the "majority functions" "2 out of 3," "2 out of 4," "3 out
of 4," and "3 out of 5" are realized. A similar investigation is
described in [7]. In [70] the majority organ used to detect and iso-
late failed elements in redundant automata is analyzed in [127].
The structure is described, and the functions realized by this type
of majority organ are presented. It is shown that its efficiency is
close to that obtained by Pierce [278]. Finally, in this group we

find the work of Polhemus [280], in which several error-correcting decoders are described. The simplest of these comprises two registers made from magnetic cores. The majority correction principle is used.

Closely allied with the foregoing is the investigation of the design of adaptive majority organs. The idea of building adaptive structures first appeared in a paper by Boshko, Glinski, and Therrien [180], in which structures with variable majority organ thresholds are described. A year later the previously cited paper by Mann [251] was published, in which techniques are considered for varying the threshold of the majority organ, along with papers by Pierce [275, 276]. In the latter a relationship is exhibited between the vote-taking principle and the principle of maximum likelihood, and it is proposed on the basis of this relationship that a decision element functioning on the threshold principle be used. The decision element estimates its input function in terms of weighting coefficients, the values of which depend on the failure probability of the corresponding input. In the event that these coefficients are determined automatically in the course of operation, the decision element is transformed into a majority organ. Pierce [275, 276] also describes several specific methods for the synthesis of adaptive majority organs and gives a comparison of the reliability of structures using conventional and adaptive majority organs.

The complexity of adaptive majority organs, however, is relatively high.

In a paper by V. I. Potapov [108] the structure of a system is implemented with threshold elements so that when the threshold is valid over a certain range the functions realized by the structure remain unchanged for all the threshold elements, regardless of the variation of the functions realized by the threshold elements themselves. It is shown that this kind of structure has an enhanced reliability. Estimates of the faultless-operation probability are obtained for 1-reliable structures. It is verified that the redundancy of the structure is lower if the threshold elements realize the "Pierce arrow function."

In [158] Tsiramua describes a method for the synthesis of discrete systems having a self-adapting structure. Its elements must be multifunctional. An example of a system consisting of bi-functional elements is discussed.

The composition of structures with adaptive majority organs has also been investigated by Domanitskii [28]. He describes a procedure for the synthesis of an optimum adaptation algorithm for redundant logical structures with variable-threshold restoring organs. The initial redundancy and adaptation algorithm is based on the condition of obtaining a probable penalty equal to or less than a given level for the output of false 0 and 1 signals from the structure. The procedure is applied to redundant structures with equal and unequal probabilities of false 0 and 1 signals at the output of the functional logic blocks. To choose the optimum algorithm it is necessary to consider the reliability of the checking and adaptation system.

Despite the fact, as already mentioned, that the majority model of the incorporation of redundancy has been the most widely used technique in practice to date, only a few papers have been devoted to its practical utilization. Blass indicates in [178] that many large equipment systems such as nuclear reactors are furnished with special "reliability-assurance" systems to protect them in the event of any deviation of operating conditions from the norm. The reliability of the safety systems is enhanced by means of vote-takers with a majority organ realizing a "2 out of 3" majority function. Moreover, the validity of the input functions is checked using informational redundancy. Fleck [200] has investigated the reliability of various majority models in order to select the one best suited to increasing the reliability characteristics of spacecraft flight-control computers. Correcting codes are used to enhance the reliability of the individual blocks. In [296] Teoste uses the majority method to increase the reliability of computers. He calculates the probabilistic characteristics of their reliable operation on the assumption that the failure probability of the elements is small and does not exceed a definite level. Similar studies have been carried out by Brown, Tierney, and Wasserman [181]. We conclude our survey of research based on the majority model with reference to the publications of Buzzel, Nutting, and Wasserman [182], Friedman [204], and Lindman [242].

§6. Methods Using the Interweaving Model

Methods for the synthesis of structures based on the interweaving model are very closely related to majority methods. The first study in which structures of this type were proposed was published, as mentioned above, by Tryon [300], who investigated redun-

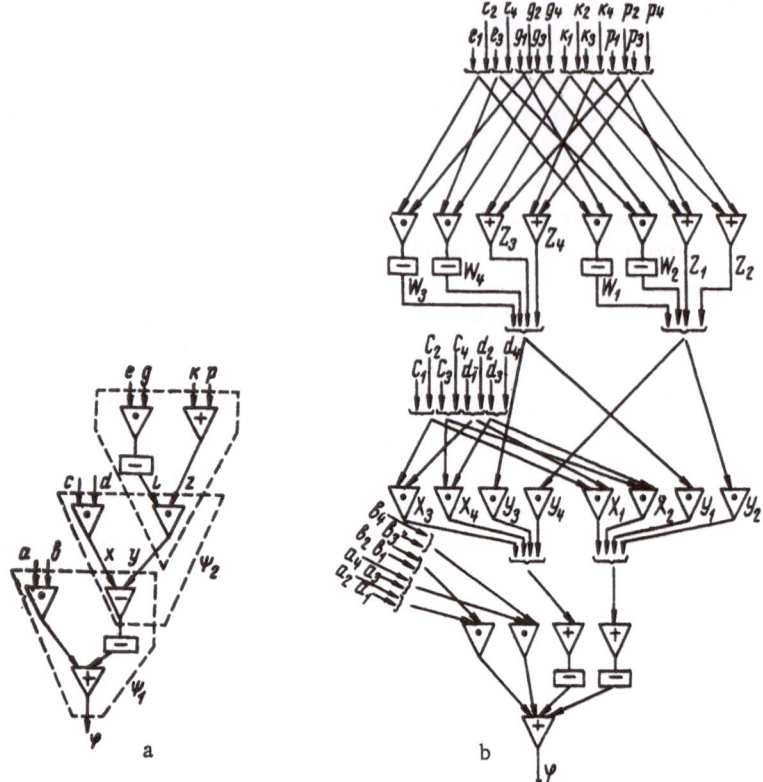

Fig. 10. Typical interweaving model. a) Primary structure; b) quadded structure.

dant structures of AND, OR, NOT elements without the use of special majority organs. Each element in the structure simultaneously realizes the logic function and correction function applied to certain failures of the preceding elements of the structure. This is accomplished by replicating each logic network a certain number of times and implementing the correction function by means of signals from blocks adjacent to the uncorrected block (see Figs. 10a and 10b, which illustrate the primary and redundant structure, respectively). When this model is used, no single failure within the structure influences the execution of the functional algorithm. Certain combinations of two or more failures are also corrected. At least fourfold redundancy is required. In [300] Tryon presents a method for the synthesis of these structures, both combinational and sequential.

The same structure has been investigated by Pierce [277], but in a more general form. He designates structures of this type as alternating-layer structures, and it is this designation that is customarily found in the literature. The idea of this reliability-improvement technique is that a failure at the output of a particular type of element due to what is for it a critical input error is a sub-critical failure for another element and vice versa. It has been verified, for example, that this property is inherent in a pair of AND and OR elements, the same pair being formed by two similar NAND elements or two similar NOR elements. The alternation of layers of such elements therefore converts critical failures in the suceeding layer into subcritical failures, and the latter do not affect the execution of the algorithm in the succeeding layer. It is assumed, of course, that the influence of failures propagates in the direction of flow of useful information (bearing in mind structures without loops or feedback links). Layer alternation in the structure can be combined with quadding.

In [278] Pierce investigates structures that annihilate the influence of failures on the proper execution of the functional algorithm (i.e., error-correcting structures) in one or two layers. In order for the influence of failure to be annihilated in one layer the failure must be subcritical, and the non-redundant prototype of the modified element must realize a logic function of more than one variable. Quadded threshold elements are described which are used for the synthesis of structures in which the influence of failure on the execution of the functional algorithm is annihilated in one layer. A constructive proof of the existence of structures with error correction by the alternation of layers is given for AND, OR, NOT elements. The possibility of synthesizing such structures in a NAND and NOR basis is demonstrated. Pierce shows that the majority organ is a special case of an alternating-layer structure and advances the hypothesis that any structure with internal error correction represents a certain type of alternating-layer structure. The possibility of estimating the faultless-operation probability of an alternating-layer structure is investigated in [278]. Error-correction methods in two successive layers are also analyzed.

A part of this group of investigations is Jensen's paper [228], in which a method is described for the synthesis of structures with alternating layers in a NOR basis and lower estimates are found for faultless-operation probability of the resulting structures.

If there are stringent constraints on the number of inputs and load capacity of the elements, it is not always advisable to use complex interwoven structures. It is possible in this case to use majority organs in conjunction with quadded elements. The properties of structures of this type and methods for their design have been investigated by Jensen in another paper [229].

The use of the interwoven-layer model for an AND, OR, NOT; NAND; and NOR basis has been explored by Lyakhovich [74]. The method for the synthesis of these structures differs from those described above in that the primary nonredundant structure is partitioned into two-layer blocks, each of which is replaced by a functionally equivalent redundant input error-correcting structure of a given multiplicity. The elementary networks of the redundant and nonredundant structures of the corresponding blocks in this case must contain similar elements. Estimates are found for the fault-less-operation probability of the resulting structures and for the domain of applicability of the method. The advantage of the method is that orthodox techniques for the synthesis of input error-correcting structures can be used for the construction of the redundant blocks. Guidelines are presented for extending the method to the synthesis of interwoven-layer structures in any basis.

Investigations involving the interwoven-layer model have also been described by Cluley [184], Youngblood and Breipohl [312], and Ichikawa and Watanabe [226].

§7. Methods Using Effective-Coding Models

7.1. Theoretical Foundations of Coding Models.
Before we begin our survey of research based on coding models, we take a brief look at some of the studies in which the problem of the efficacy of redundancy for the reliability improvement of discrete systems has been investigated.† Shannon [159] proved the fundamental coding theorem for a noisy discrete communication channel, exposing the conditions under which the reliable transmission of messages via such a channel is possible. The theorem was proved in two stages. It was shown in the first stage that it is possible to

---

†This problem is discussed in the present section, because its investigation was begun after the publication of von Neumann's paper [269] and Moore and Shannon's paper [263] and is based on their results.

construct for every discrete communication system an encoder that will translate an input signal into a sequence of binary symbols at a mean rate R (the transmission rate is defined as the ratio of the number of information symbols to the total number of symbols in the encoded word, i.e., $R = k/n$, where k is the number of encoded symbols and n is the number of symbols obtained after coding), plus a decoder that translates this sequence of binary symbols into a corresponding replica of the primary signal. It is proved in the second stage that it is possible to determine for the channel its carrying capacity C (in binary symbols per second: bits/sec), i.e., the maximum mean rate at which reliable transmission via the channel is still possible. The substance of the theorem is that the input can be restored with a certain reliability from the output if the transmission rate R is less than the carrying capacity C. In this case two independent quantities R and C are compared, the first of which characterizes only the input–output pair of the channel, while the second characterizes the channel itself. The transmission errors per se do not limit the reliability, but only the carrying capacity C. Given a fixed transmission rate R and carrying capacity C, the reliability can only be increased by increasing the complexity of the encoding and decoding devices.

In [196] Elias proposes the investigation of a combinational discrete control system as the analog of a discrete communication system† in which information is not only transmitted, but is transformed as well. Elias represents this system as an absolutely reliable combinational system in series with a communication channel subject to noise. The reliability of the system is increased by the insertion of encoders and decoders, which reconstruct the true values of the outputs from the set of output states. Thus, only informational redundancy is introduced, the additional structural redundancy occurring merely as a result of the informational redundancy. Elias [196] and, after him, Cowan [188] sought to prove the analog of Shannon's theorem for C–systems by investigating Boolean functions of two variables; they found that of sixteen possible functions only the eight even–numbered conjunctions enter into the perfect disjunctive normal form as disjunctive terms. These functions

---

†Systems of this type are called computing systems by Elias [196], Cowan [188], Winograd [308-310], Glinski [17], and others. We shall refer to them hereinafter as C–systems.

include the following:

$$0, 1, x_1, x_2, \bar{x}_1, \bar{x}_2, x_1 \oplus x_2, \overline{x_1 \oplus x_2}. \tag{3}$$

These functions are called "even" functions [196]. It was verified as a result of the investigations that the use of a correcting code yields a greater-than-zero information transmission rate through the C-system for vanishingly small error frequencies if and only if the C-system realizes even functions. Of the eight possible even functions (3), however, only two are essentially dependent on two variables: $x_1 \oplus x_2$ and $\overline{x_1 \oplus x_2}$. It is readily perceived that they are not universal, i.e., they do not form a basis from which it would be possible to construct an arbitrary Boolean function. Elias [196] showed that the best code for odd functions in the redundancy technique described above is one such that positional encoders repeat every value n times and accept the valuation of the majority of outputs as the correct one. This means that an arbitrarily high transmission reliability in the C-system for these eight functions can only be achieved at the expense of vanishing information processing rates. Elias' analysis [196] was repeated by Peterson and Rabin [273], who arrived at essentially the same result.

In other words, with the introduction of informational re-dundancy† the error probability tends to zero together with the rate. The investigations of Löfgren [71, 72] produced a similar result. Cowan [188] and Winograd [308, 309] investigated the method of von Neumann from the same point of view. Von Neumann's scheme uses structural redundancy, because an element with a single line is replaced by n elements with a bundle of n lines. In this scheme the bundle carries one information bit, so that the rate of processing of the latter is equal to 1/n bits (but per line, rather than per symbol). Consequently, if structural redundancy is used, absolute reliability can only be achieved at zero rate.

In [188] Cowan proposes the use of more complex elements in C-systems, so that the restoration of information is realized along with the realization of the Boolean operation. The result of this notion, however, is merely that the rate still tends, though more slowly than in the foregoing models, to zero as the faultless-operation probability is increased.

---

†In the sense implied above; i.e., the structural redundancy used is merely a consequence of the introduction of informational redundancy.

Subsequent investigations by Cowan and Winograd [311] showed, however, that both types of redundancy, i.e., informational and structural, can be used to render the discrete system as reliable as desired. They proved the analog of the fundamental theorem of Shannon in application to C-systems. Naturally they exclude failures inherent in the outputs, because Rabin established the existence of a class of automata that cannot execute the functional algorithm at arbitrarily small error frequencies without the use of reliable output elements. The redundant system whose feasibility was proved by Cowan and Winograd [311] must have a structure such that the initial functions are computed in several locations and any element of the structure realizes a complex Boolean function that is the composition of many initial functions. The mean number θ of inputs per element for a d-reliable system in this case is limited by the expression

$$\theta \geqslant (2d + 1) \frac{k}{n},$$

where k is the number of information symbols of the particular correcting code and n is the word length of the correcting code.

The necessity of increasing the complexity of the elements in order to assure the d-reliability of a structure has been demonstrated by Löfgren [71]. The error-correcting capacity of discrete systems has also been investigated by Harrison [216].

7.2. Methods Using the Input-Correction Model. The methods based on this model are few in number. We have already cited Peterson and Rabin's paper [273], in which they investigate two modifications of the model: iterated positional computation with subsequent decoding and encoding of the inputs by an effective correcting code using checking and error-correcting devices. In the same vein is a paper by Winograd [310]. He considers only synchronous automata, which are insensitive to transient errors. The necessary and sufficient conditions for the existence of this type of automaton call for the existence of a finite sequence to function as a universal synchronizing signal. A relationship is found between these automata and variable-length codes. Aspects of automata that are "self-adapting" to correct operation after a transient failure have also been investigated by Levenshtein [65] and Zarovnyi [44].

7.3. Methods Using the Memory-Correction Models. As we mentioned earlier, this model was first intro-

duced by Gavrilov [9] for the correction of failures in memory
elements. In the case of contact relay devices, in which a failure
of the system logic block can be related to failure of the corresponding memory element, encoding of the states of the memory
elements ensures the same degree of reliability of the logic
generator (provided the latter realizes functions y_i and z_i in the
minimum disjunctive normal form), because the introduction of
redundancy into the memory block automatically produces redundancy in the logic block.

In Gavrilov's paper [10], which is a continuation of [9], he
shows that after the introduction of redundancy into the memory
block the functions y_i and z_i are transformed into generalized
symmetric Boolean functions.† He also indicates how the signalization of memory element failures can be implemented for this
redundancy model.

Krishtal' [60] has used the same redundancy principle with
a code that corrects d errors. However, for any representation of
the number d as the sum of two nonnegative numbers, $d = d_0 + d_1$,
the structure of the logic block of a relay contact system is realized so that it becomes (d_0, d_1)-reliable, where either a failure of
one relay coil (and, accordingly, a like failure of all its closing
contacts and failure of the opposite type of all its opening contacts)
or the failure of any one contact is interpreted as a single failure.

The same model of a redundant device has been used in
studies by Karpovskii [49-51] for the synthesis of d-reliable automata. The states of the internal elements are coded with regard
for the input states. Inasmuch as the latter are assumed to be reliable, their inclusion makes it possible to diminish structural redundancy. Necessary and sufficient conditions are proved for the
existence of a d-reliable automaton.

In [30] Doncheva uses the same encoding principle as in [9].
The logic block in this case is simply based on the division-of-
states method often used in the theory of relay devices for the purpose of simplifying the structure.

A similar method for the simplification of reliable structures
has been employed by Turuta [151] as well as Lazarev, Piil', and

---

† The generalized symmetric Boolean function is defined as a function that trans-
forms into a symmetric Boolean function under a certain change of variables.

Turuta [63]. Moreover, simplification is achieved by the use of threshold elements. The specific characteristics of a particular functional algorithm as reflected in a special form of distribution of the probabilities of its internal states and different hazards of dissimilar failures are taken into account in [149].

In two other papers by the same author [148, 150) the method of coding vectors $(y_1, \ldots y_r)$ by a d-error-correcting code is extended to structures synthesized from pulse elements. It is shown that the properties of these structures can be taken into account to reduce the redundancy required for the assurance of d-reliability. Specifically, after encoding, certain internal states can be separated by a distance less than 2d + 1.

Gridin [23] has extended the method to the case in which diodes are used to realize the logic block.

Also in this group of papers are the investigations of Frantsis and Yanbykh [155] and Frank and Yau [202]. In these papers, however, the correcting block is not co-existent with the main part of the device. In the first paper it is constructed from magnetic elements. In the second paper a corrector is used to correct asymmetric failures.

In [281] Rao investigates the reliability gain afforded by the introduction of redundancy through the assignment of a correcting code to the states of memory elements. It is noted that the introduction of redundancy has both a positive and a negative effect on the reliability of the system in that, on the one hand, the system acquires a higher degree of reliability and has the capability of more rapid error detection, but, on the other hand, the number of elements in the system is increased, the redundant elements are subject to failures as well, and this fact can tend to lower the reliability. The author's analysis shows that the net effect of the encoding of states on the system reliability is positive, i.e., the introduction of redundancy by means of error-correcting codes enhances the system reliability if the number of additional logic elements and memory elements does not exceed a certain fraction of the total equipment volume of the primary system. This fraction depends on the extent to which the particular checking system used augments the reliability of the system.

We should also mention Goldman's paper [209], in which a

cyclic code is described for increasing the reliability of a discrete-system memory block assembled from magnetic cores.

7.4. Methods Using the Logic-Correction Model.

As we mentioned above, this model was first proposed by Elias [196] for the correction of failures in logic elements. Informational redundancy is introduced in this investigation, and the resulting structural redundancy is merely a consequence of the informational variety. The central problem studied in this paper is the propriety of using code assignments to improve the reliability of combinational devices in the example of two-input systems. No mention is made of any specific realization of the model.

A paper was published a year later by Zakrevskii [35], who introduced structural redundancy on the basis of the assignment of Hamming codes to the values of the l functions realized by l outputs of a combinational discrete system. The encoding operation produces a certain number h of checking functions, whose dependence on the primary functions is determined by the form of the code used. Prior to realization of the checking functions the values of the primary functions are inserted into the expressions for the checking functions, yielding a system of l + h Boolean functions of the same input variables. Decoding is effected by a corrector having the same structure as required for decoding of the correcting code. In the case of a single-output system the functions presented to the inputs of the system output element are coded. This method for the introduction of redundancy is elaborated in more detailed form in two other papers by Zakrevskii [36, 37].

Investigations similar to those of Elias [196] are described in a paper by Winograd [308], in which it is shown that of the total 2^{2^n} possible Boolean functions of n variables only $2^n + 1$ functions are realizable at nonzero information-processing rates under conditions analogous to those assumed by Elias in [196]. These functions also fail to form a basis, i.e., they do not contain the elementary functions required for the realization of the other $2^{2^n} - 2^n - 1$ functions.

These studies are continued in another paper by Winograd [309]. He investigates the reliability enhancement of combinational discrete systems containing unilateral conducting elements and multivalued elements. He uses the same redundancy model as in [308]. Failures localized in the encoders and decoders are not taken into account. A lower estimate is derived for the complexity of an element of a d-reliable automaton in the form $\theta \geq (2d + 1)k/n$, where k is the number of information symbols and n is the total number of symbols in the words of the d-error-correcting code; the complexity is expressed in the number of inputs. Consequently, the higher the degree of reliability desired, the greater must be the complexity of the element.

Subsequent results in this direction are found in Winograd and Cowan's book [311]. We have already mentioned the proof in this book of the analog of Shannon's theorem for noisy communication systems in the case of discrete control systems. According to the analogous theorem, an automaton can realize any Boolean function with arbitrarily high reliability at a nonzero information-processing rate in the span of a certain fixed time interval. This result was obtained by modification of the model treated by Elias (which is essentially the analog of the model used by Shannon to investigate information transmission systems). The distinctive feature of Winograd's model is the simultaneous introduction of informational and structural redundancy. In this sense the study is somewhat reminiscent of an earlier investigation by Eden [194], in which a single failure is detected at nonzero information-processing rates, where a portion of the system operations are executed by the decoder.

Failures in the encoders and decoders (other than failures in the output circuits) and failures in their interconnecting elements are also included in Winograd and Cowan's book [311]. The required redundancy in this case is smaller, the greater the number of element inputs. In order to obtain a single-cycle discrete system with a failure probability not greater than Q (Q $\geq$ 0) elements are used with a failure probability q satisfying the relation

$$q \leqslant 1 - (1 - Q)^{M-1}, \tag{4}$$

where M is the number of elements of the primary system. It is assumed that the elements from which it is required to realize the system have a definite carrying capacity C that is independent of their complexity.† It is shown that for sufficiently large k there exists at least one (n, k)-code such that $q \cong 2^{k-nC}$. There-

---

† If C varies with the complexity (i.e., with the number of inputs) of the element, the redundancy-introduction procedure is complicated somewhat, but exactly what is involved is not elucidated in [311].

fore, for any finite M and any given Q relation (4) can be satisfied at the expense of an increase in the code parameters k and n such that

$$\frac{k}{n} < C.$$

We now describe the Winograd — Cowan technique for the introduction of redundancy [311]. We assume for simplicity in the beginning that failures at the output of the primary structure are due solely to failures of its elements and not of their interconnections. Let the number Q representing the desired failure probability of the synthesized system containing M elements x_1, ..., x_M be given. Since the failure of these elements (rather than their inputs) are responsible for failures of the system, their states are assigned codes. Suppose than an (n, k)-code is used. The parameters n and k are chosen so as to make the carrying capacity of the noisy communication channel for this code equal to C and to make the code have an incorrect-decoding probability smaller than or equal to Q. The selected code determines the encoding and decoding functions. Before coding is effected the primary system is replaced by a set of k replicas of that system. As a result we have in place of each element x_i a set of k elements x_{i_k}, ..., x_{i_k} realizing the same Boolean function f_i k times. The selected (n, k)-code is then applied to these elements, i.e., encoding and decoding are realized simultaneously for each of the M sets of k similar elements, and if these elements are input (output) elements, they are replaced by elements which, in addition to the prescribed functions, also execute encoding (decoding), but if these elements are internal, they are replaced by elements which, in addition to the prescribed functions, execute decoding of the information transmitted to them and encoding of the information extracted from them. Failures generated by false connections between elements are treated as additional errors in the code word, and their influence is annihilated by an appropriate increase in the code parameters.

Among this group of papers are two by Löfgren [71, 72], in which the efficacy of self-organizing systems in the sense of prolongation of their service life is demonstrated. Such systems automatically replace incorrectly functioning elements or use self-restoring systems.

A study by Gore [211] is similar to the work of Elias in the redundancy model used. It is proposed that the primary structure be realized as a two-layer (two-level) system. In this case the failure of elements of the first layer (level) may be interpreted as an input error for elements of the second layer. Consequently, the input states of the second layer (M in number) are encoded. According to the fundamental coding theorem there is a number N of outputs of the first layer such that at least 2^{R^N} distinct input states can be restored after noise effects with a failure probability $Q \leq 2^{-(c-R)N}$ for the channel connecting the two levels, where C is the carrying capacity, $C = 1 - p \log_2 p + q \log_2 q$, q is the element probability, $p = 1 - q$, and R is the information-processing rate, $0 < R < C$, $R = M/N$. The redundancy used here is $I = N/M$, where M is the number of second-level inputs. This structure is realized h times for the correction of failures of second-level elements, where h is determined with regard for the cost of the system as a function of the cost C_0 per binary input and the total redundancy. The upper bound N_u of N minimizing the average cost

$$C_a, \left(C_a = \frac{C_0 hN}{(1-Q_e)^h} \right),$$

which is determined by the expression

$$2^{CN_u} = 2^M (1 + hCN_u \ln 2)$$

or, with regard for the value of Q_e,

$$N_u = \frac{1}{C} \{ M + \log_2 (h \ln 2) + \log_2 [M + \log_2 h \ln 2 + \dots] \}$$

is determined.

The redundancy model used by Tooley [298] is similar to the one used by Elias. Let the primary system have m inputs and realize l Boolean output functions. As in Elias' case [196], the starting point is a C-system with symmetric failures. This concept justifies the assignment of error-correcting codes to the l functions realized by the outputs of the primary system. Every state a corresponding to a code word $\alpha = (\alpha_1, \dots, \alpha_{l+h})$ is treated as a point in n-dimensional Euclidean space with coordinates α_1, ..., $\alpha_{l+h} (n = l + h)$. The set of all states into which the state a is thrown by failures of any t ($0 < t \leq d$) elements, where that set is treated as a set of points in n-dimensional Euclidean space, lies inside (inclusive of the boundary) a sphere described from the point α as its center. The output states are decoded by the logic block on the principle of "condensation" of states corresponding to points of spheres with integer coordinates toward the centers of the corresponding spheres. Thus, the logic block maps all 2^{l+h} realizable sets of values of the l + h functions onto 2^l output states. This mapping is determined by the selected correcting code. The coefficient I, defined as the ratio of the failure probabilities of the redundant to the primary system, is adopted as the criterion of the gain in system reliability. It is noted that the inputs of the primary system are sometimes the outputs of other failure-prone systems. In this case encoding is applied to the output functions of both systems, while decoding of the outputs of the first system is transferred into the second system. Consequently, the functions realized by the given system (2^{l+h} in number) are not functions of m, but of m + k arguments. The realization of these 2^{l+h} functions is unconnected. The behavior of the coefficient I when various codes are used to correct the same number d of errors is analyzed in the same paper.

Swoboda [295] has investigated the detection of failures in a logic converter. He uses a network that differs from the one shown in Fig. 6 only insofar as the correcting device is replaced by a signalizer, which generates the values of checking functions at the logic block outputs and compares them with the analogous values generated by the encoding block. In the event they fail to coincide, the signalizer generates a failure signal. It is tacitly assumed that the auxiliary blocks have a much greater reliability than the primary system, or that the probability of the simultaneous occurrence of more than one failure is negligible. Even with the occurrence of one failure the values of the functions realized at several outputs of the logic block can be distorted. Consequently, before the code is selected the maximum number d of simultaneously impaired outputs is determined. A linear code correcting d errors is used. The mathematical foundation of this redundancy technique may be found in a paper by Kämmerer [230].

The effort to increase reliability at the expense of less redundancy introduced into the structure led to utilization of the internal redundancy present in a system. This problem has been solved in only a few papers. Typical representatives are [102, 74, 37].

In [102] Petrosyan investigates single-output discrete systems. The set of all sets of values of the input variables is

partitioned into two subsets M_0 and M_1 depending on the corresponding output value of the system. The reliability of the system
is increased by encoding of the input variables. It is noted that a
change in the output occurs only when an episodic failure can be
matched against a change of state from M_0 to M_1 or from M_1 to M_0.
It is proved on this basis that it suffices for the synthesis of this
type of d-reliable system to ensure a distance $D = 2d + 1$, not
between any pair of input states, but only between pairs of states
belonging to different sets M_i ($i = 0, 1$). This result has been
generalized by Lyakhovich [74] to the case of multi-output structures. In this case the number of distinct sets M_i is equal to the
number of distinct groups of functions realized by the primary
system outputs. Internal redundancy has also been used by Zakrevskii [37], who describes a method for the synthesis of $(0,d_1)$-relia-
ble contact structures realizing incompletely determined Boolean
functions. We note that only one type of failure is considered in
this paper.

Concurrently with [37] a number of papers appeared in
which methods were investigated for the improvement of reliability
with regard for the types of failures involved, i.e., on the assump-
tion that the probabilities of $0 \rightarrow 1$ and $1 \rightarrow 0$ failures had different
values (sometimes, as in [37] for example, it is assumed that the
probability of a particular type of failure is so close to zero as to
render it negligible).

Representative of this group of papers are [282, 46, 125].
In [282] Rau considers d-reliable systems composed of n elements.
The probabilities p_1 of $0 \rightarrow 1$ failure and p_2 of $1 \rightarrow 0$ failure of
the individual element are assumed to be known. The limiting
values of d ($d = [\tilde{d}] + 1$) for given n, p_1, and p_2 as well as the
limiting values of n ($n = [\tilde{n}] + 1$) for given d_1, p_1, and p_2 are
determined, where

$$\tilde{d} = n \, \frac{\log(1-p_2) - \log p_1}{\log[(1-p_1)(1-p_2)] - \log p_1 p_2} \, ,$$

$$\tilde{n} = d \, \frac{\log p_1 p_2 - \log[(1-p_1)(1-p_2)]}{\log p_1 - \log(1-p_2)} - 1.$$

In [46] Ivas'kiv and Ryakin determine the condition that
must be met by the structure of a combinational system in order
for its outputs to be endowed with ultimate asymmetric failures
(i.e., failures for which either p_1 or p_2 vanishes). The primary

system is regarded as a C-system with an asymmetric communication channel. It is proposed for the correction of asymmetric failures that an asymmetric-error-correcting code be used. These investigations are continued in Ryakin's paper [125]. He considers the application of a definite class of correcting codes to check unit failures in structures assembled from NAND, NOR, and AND, OR, NOT elements.

A separate group of papers is concerned with the enhancement of reliability through the use of special classes of codes. Palounek [272], Roy-Chaudhuri [286], Frantsis [154], and Frantsis and Yanbykh [156], for example, investigate the possibility of using correcting codes with a base b > 2. The first-published paper was by Roy-Chaudhuri [286]. He augments an l -output primary combinational system with k unconnected realizable subsystems, each of which has an equal number l/k of outputs. A method is proposed for the synthesis of a code correcting d = 1 error with base b = $2^{l/k}$, which is capable of encoding the sets of values of all l functions. Each single failure of an i-th symbol in the code words corresponds to one and only one combination of distorted values t $(0 < t \leq l/k)$ of the outputs of the i-th subsystem. In Frantsis and Yanbykh's paper [156] as well as in Palounek's paper [272] other methods are described for the synthesis of such codes, and estimates are given for their parameters when d > 1. Khristal' and Ostianu [62] partition the primary combinational system into g subsystems, each i-th member of which realizes h_i outputs. The values of h_i can differ for different values of i. It is shown that the assignment of a code to the sets of output values of the system can be used in combination with pulse designators correcting d errors for the synthesis of d-reliable structures in which the distortion of t outputs of one of the subsystems, say the i-th (in this case $0 < t \leq h_i$) is interpreted as a single failure.

The basis for the partition into subblocks is served by the following considerations. One given subblock must realize those functions for which the failure probabilities of its corresponding outputs have the same order (assuming that the failure probability is specified for each output of the system). If the structure of the primary system cannot be readjusted, it is regarded as the one and only subblock of the partition. In this case the code contains just one information symbol, which is multivalued. As a result the structure can be realized without invocation of the majority prin-

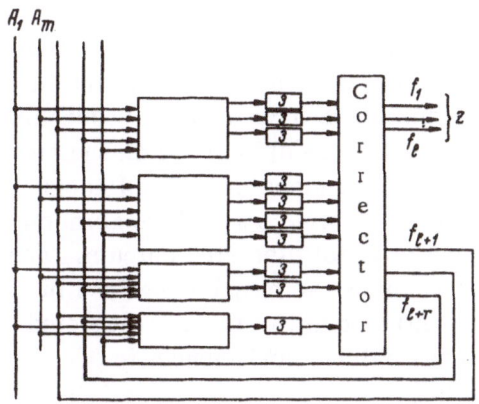

Fig. 11. Redundancy model using
a correcting code in combination
with pulse symbols.

ciple. A procedure is described for the synthesis of a d–reliable
corrector for this class of codes. It is noted that the given procedure can be extended to the synthesis of d–reliable sequential systems. The network of a typical structure is illustrated in Fig. 11.
The figure portrays a redundant system with feedback, because,
generally speaking, this method can be used without any special
difficulty to increase the reliability of sequential systems. In [93]
Ostianu considers some estimates of the faultless–operation probability of this type of d–reliable structure.

Structures that are reliable with respect to a specific list
of system output failures have been investigated in papers by
Savchenko [126, 128] and Zarenin [42] as well as in a joint paper
by these authors [43]. Redundancy is introduced by the network
illustrated in Fig. 12. The structure of the primary system is left
unaltered in this case. A procedure is described for the synthesis
of a special code to correct the error lists.

In Ryakin's paper [124] the primary device is also left unchanged, but the auxiliary redundancy is introduced only for the
detection of individual failures inside the structure resulting in
dependent output failures. Failures are located by means of a code

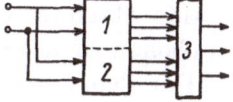

Fig. 12. Redundancy model correcting a
specified list of failures.

belonging to one of two classes of error-correcting codes. The
first class of codes detects independent d-fold errors at the outputs
of either the primary system or the auxiliary system realizing the
checking functions. The codes of this class have a lower redundancy
than the conventional codes used to detect d-fold errors on the entire length of the code words. The second class of codes detects
independent batches of errors of length t, where t is the maximum
number of outputs between the first and last distorted outputs. The
primary and auxiliary structures in this case may be realized in
the connected mode. The primary structure can be reconstructed
in order to reduce the length t of the batches and, hence, to decrease the redundancy of the code.

Ostianu, Lyakhovich, and Potekhin [95] have developed a
new class of linear correcting codes, which they call "duplicating"
codes, in order to decrease the complexity of a redundant structure.
The codes are so named because the number h of check symbols
used in them for d = 1 is equal to the number of code information
symbols. It is shown that these codes yield a corrector complexity
(expressed in the number of implicants) equal to $8l$ and lower than
the complexity of correctors for minimum-redundancy linear
codes. The complexity of the rest of the redundant structure is
lower than attained by triplication. The conditions are determined
under which the code is more effective in terms of complexity than
triplication, viz.:

$$s_0 > \frac{s_1 + \Delta s + s_l}{3},$$

where s_0 is the complexity of the primary structure, Δs is the
additional structural redundancy generated by the unconnected
realization of the primary structure, s_1 is the complexity of the
unconnected realization of the checking functions, and s_l is the
difference between the realization complexities of the l-output
corrector and l single-output majority organs.

In [271] Ostainu proposes the encoding of output states
of the primary system by means of a correcting code, each code
word of which represents a sequence of μ code words of different
d-error-correcting codes. Each part of the code word representing a code word of the indicated μ codes is compared with the
outputs of one and only one of the μ subblocks into which the primary system is partitioned. The number μ is chosen so as to
make the failure probability of each output smaller than a speci-

fied quantity Q_i $(0 \le Q_i < 1; i = 1, ..., l)$. The code is decoded by means of μ correctors; the functions decoded in different correctors are realized in the connected mode. From this method of synthesis the author develops as special cases a method using one corrector ($\mu = 1$) and a majority method ($\mu = m$). A block diagram of the redundant structure obtained by the method described is shown in Fig. 13.

We also examine here two papers devoted to comparisons of some of the reliability-improvement methods discussed in the present section. One is a paper by Sagalovich [130], and the other is by Akers [291]. In [130] three methods for the synthesis of reliable contact systems are compared in terms of complexity and the probability bounds of $0 \rightarrow 1$ and type $1 \rightarrow 0$ failures. They are the method of Moore and Shannon [263], the method of minimal bracketed forms described by Krishtal' [57], and the method of symmetric lattices proposed by Sagalovich [130]. The potential practical applications of the Moore — Shannon [263] and Winograd — Cowan [311] methods are discussed in [291], along with the behavior of the operational reliability characteristics of so-called "polyfunctional nets." Two other papers dealing with the existence of arbitrarily reliable discrete system structures follow the same vein. They are by Muchnik and Gindikin [183, 16] and Kirienko [55]. The following problem is treated in [83]. A system of Boolean functions is given, consisting of two subsystems A and B, where the functions from A are realized by absolutely reliable elements

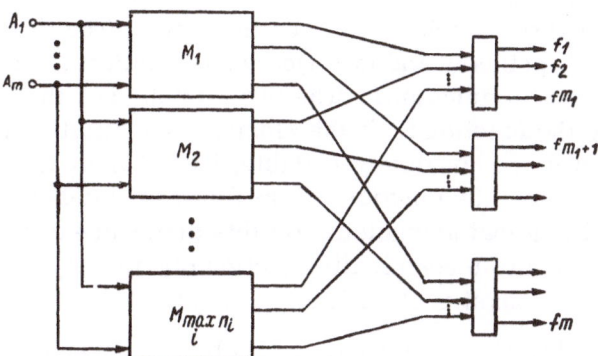

Fig. 13. Redundant structure model using several correctors simultaneously.

(with a failure probability of zero), while the functions of B are realized by elements characterized by known upper limits of the failure probabilities. The conditions are determined that must be satisfied by the system of functions A ∪ B, along with the form of the functions that each subsystem must contain in order for any Boolean function to be realized from the elements realizing all functions from A ∪ B, with a failure probability smaller than any predetermined number Q (Q > 0). A system A–B admitting an arbitrarily reliable realization of any Boolean function is said to be h–complete. Necessary and sufficient conditions for the h–completeness of a system of functions are deduced.

Very closely allied to [83] in terms of the object of investigation is Kirienko's paper [55]. He introduces the concept of the d–completeness of a system of functions C as the property of realizing any Boolean function of a d–reliable structure synthesized from elements realizing functions from C. Necessary and sufficient conditions for the d–completeness of a system of functions are determined. Also, the realization complexity of any Boolean function of n variables $f(x_1, ..., x_n)$ by d–reliable structures synthesized from functional elements is investigated. The function to be realized is decomposed into the last t variables, whereupon the function f is written in the form

$$f = F(f_1, ..., f_{2^t}),$$

where $f_i = f_i(x_1, ..., x_{n-t})$ (i = 1, ..., 2^t) are functions of the first n − t variables. The resulting 2^t functions are assigned a linear correcting code, which corrects d errors and has 2^t information and (t + 1)d check symbols. The function f is realized d–reliably in the form of three blocks connected in series. The first block realizes the 2^l primary and (t + 1)d check functions. The second block decodes these functions. The third block uses the values of f_i to restore the function f. If the elements constituting the second and third blocks are absolutely reliable, i.e., if their failure probabilities are zero, the function f is realized by a d–reliable structure. It is shown that asymptotically this structure does not increase the realization complexity function obtained by Shannon [159] for nonredundant realizations.

7.5. Methods Using a Hybrid Correction Model. We have already indicated the first paper published in this area, by Armstrong [167], in which two separate correctors

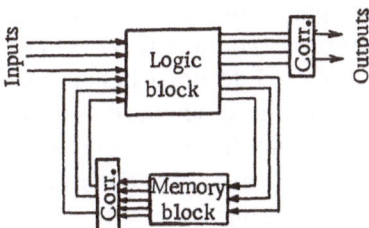

Fig. 14. Logic- and memory-correction
model.

are used to correct failures in the memory block and failures in
the logic block (see Fig. 14), since the output functions and feedback functions are assigned independent correcting codes. The
average faultless-operation times T_1 and T_2 are determined for
each corrector. Each corrector is executed in two copies. For
the elimination of corrector-generated failures the correctors
are replaced by their duplicates after time intervals $t_1 < T_1$ and
$t_2 < T_2$. While in the inoperative state the correctors are sub-
jected to preventive maintenance. The magnitude of the resulting
redundancy is discussed, although precise estimates are not
obtained.

The same notion is used in a study by Frantsis and Yanbykh
[155] for the synthesis of d-reliable automata. The synthesis of
an automaton is effected by the method of delay subnetworks.

Zakrevskii [39] has investigated a reliability-improvement
method based on two correcting codes for the separate encoding
of input and output states.

Krishtal' [57] has investigated the synthesis of relay contact and trigger-diode structures that are d-reliable with respect
to the inputs and internal elements (elements of the logic converter
as well as elements of the memory block). This result is achieved
by a correcting code with m + r information symbols, which are
used to encode the states of all inputs of the logic generator, external and internal. Decoding is allocated to the logic generator. For
the assurance of d-reliability of the logic block the functions realized by it are constructed in the minimal disjunctive normal form
(MDNF) or bracketed form derived from the MDNF, and the elements of the last level are d-fold replicated. (The block diagram
of the redundant structure differs from Fig. 8 only in the number
of inputs, which is equal to m.)

This procedure is extended in a paper by Ostinau and Krishtal' [94] to the synthesis of (d_0, d_1)-reliable structures. A correcting code with a coding interval $D = d_0 + d_1 + 1$ is used here. In another paper by Krishtal' and Ostianu [61] the corrector method described in [35] is generalized for the synthesis of d-reliable (with respect to internal elements) sequential systems constructed from functional elements. A d-error-correcting code with $l + r$ information symbols is assigned to all outputs of the logic block (external and internal). The primary and checking functions are realized unconnectedly by single-output blocks. The outputs of the latter are transmitted with delays to the corrector inputs. The restored output functions z_i and feedback functions y_i are taken from the corrector outputs. These functions represent the restored internal inputs of the logic generator. The corrector realizes the functions unconnectedly in the MDNF, and the elements of the last layer are d-fold replicated, either in series or in parallel, depending on the properties of the particular elements used. Consequently, the structure of the corrector also turns out to be d-reliable. A block diagram of a system realized by this method is presented in Fig. 15.

In [59] Krishtal' consolidates the methods described in [61] and [57] in order to synthesize functional-element structures that are d-reliable with respect to the inputs of the memory and logic elements. Both external and internal auxiliary inputs are required. A block diagram of the redundant structure is presented in Fig. 8.

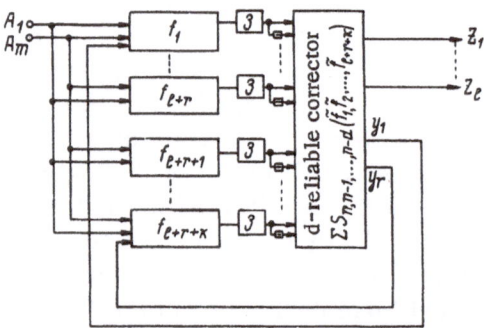

Fig. 15. Logic- and memory-correction model for the realization of a structure from functional elements.

Gavrilov, Krishtal', and Ostianu [208] have generalized the procedure described in [59] to the synthesis of (d_0, d_1)-reliable structures of functional elements (the block diagram given in Fig. 15 remains unaltered in this case).

The model described in the above-cited book of Winograd and Cowan [311] (see 7.4) for the introduction of redundancy into combinational systems is also used for sequential systems without any changes, except that in this case the failure probability q of a memory element must satisfy the relation

$$q \leqslant 1 - (1 - Q)^{b-1},$$

where Q is the given failure probability of the final system and $b = \sum_{\nu=1}^{M} a_\nu$. Here M is the number of elements of the primary structure, and a_ν is the number of cycles in which the delay of the memory element is operative.

In [134] Sagalovich investigates two well-known techniques for the introduction of redundancy into the memory block of an automaton in two realizations: through contacts and functional elements. The former is used whenever both the restoring and memory elements are subject to failures. The latter is used in the event only memory elements are failure-prone. Accordingly, the code used to correct d errors is assigned so that the minimum Hamming distance is equal to 2d + 1 between complete states of the automaton in the first case, while in the second it is equal to that between internal states only. It is assumed that the functional elements and contacts are absolutely reliable. It is shown that the complexity of the realization of a Boolean function by the combination block of the automaton increases linearly as the redundancy in the case of structures of functional elements, whereas in the contact case it increases as the three-halves power of the redundancy. Given two automata comprising the same number of elements, the less complex will be the combination block of the automaton that has the greater degree of reliability on the part of its memory elements. The advantages and disadvantages of using two different types of corrector structures are argued (the types of corrector structures will be discussed in 7.6).

Also pertinent to the present section are the results described in the previously mentioned paper of Swoboda [295]. The detection of failures in a sequential system as describes in [295] is realized in the same way as in a single-cycle system (see 7.4), because the system is treated in each cycle as a single-cycle system (i.e., it is represented as a single-cycle equivalent). The probability of an undetected failure in one cycle is estimated.

Further investigations using the same technique for the introduction of redundancy have been conducted by Kämmerer [231]. He gives a mathematical justification for the fact that a nonredundant automaton with zero reliability can be expanded to an automaton with nonzero reliability (the output failures, of course, are not corrected).

7.6. Investigation of Corrector Structures.
Considering the vital role that the correctors play in the enhancement of discrete system reliability, a number of authors have investigated techniques for the synthesis of correctors from various elements, their reliability characteristics, and their realization complexity.†

One of the first papers on this topic was written by Gavrilov, Ostianu, Rodin, and Timofeev [13], who described general principles for the construction of correctors and gave specific realizations in the contact variant, along with those constructed from electronic elements or magnetic elements with a rectangular hysteresis loop. These correctors were constructed for use in communication systems, but they can be used with equal success in discrete control systems, but they can be used with equal success in discrete control systems, because they are parallel rather than series processors. All of these correctors are synthesized on the "condensation" principle. This means that the inputs are all states corresponding to points of n-dimensional Euclidean space with non-integer coordinates inside (including the boundary) spheres of radius d whose centers are points corresponding to the code words. The states corresponding to the centers of these spheres are realized at the outputs.

Of the same general type are the correctors described by Tsandekh [313], which have a matrix structure composed of resis-

---

† We shall not be concerned in this section with the structure of devices used to decode the correcting codes and designed for operation only in discrete communication systems. The reader can obtain exhaustive information on these decoders from Gurvit's book [24].

tances and transistors. In [313] the faultless–operation probability
of a corrector thus synthesized is assessed. The correctors des–
cribed in the paper are designed for error correction in counters.

The structure of a corrector that realizes output functions
in the unconnected mode in the MDNF is described in the previously
cited papers [59, 62]. The elements of the last layer are replicated
in series or parallel, depending on the properties of the elements
used. As a result of this mode of synthesis of the corrector, its
structure does not decrease the reliability of the structure as a
whole. A corrector structure is developed in [59, 62] for codes
correcting symmetric and asymmetric errors as well as codes with
the combined use of pulse designators. In another paper [58] Krishtal'
proposes a reliable corrector structure for codes correcting error
batches. Structures of the same type are invested in the correctors
described by Doncheva [31], which are capable of correcting errors
with regard for their type. Also, any unused states are taken into
account in obtaining the output functions of the corrector.

Kautz [232] has noted that a decoder with sequential information processing can always be converted so as to be capable of processing information in parallel. A greatly complicated iterative
structure is the result. It is proposed for simplification of the
corrector structure that correcting codes with a low parity–check
density (described in [106]) be used. A corrector of this type, which
can be realized from various types of elements, including elements
of the classical basis "sum mod 2" and threshold elements, is described in [232]. A somewhat different structure is acquired by
correctors realizing a sequence of operations for the reproduction
of information with decoding of the correcting codes. In this case
the corrector K represents a combination of three devices: K',
which computes the so–called syndrome†; K", which computes the
error vector‡; and K‴, which sums the error vector with the vec-

---

† In the theory of error-correcting codes (see, e.g., Peterson [106]) the syndrome is
  defined as a vector S of length n - k (where n is the length of code words of the
  correcting code and k is the number of its information symbols) formed as the product
  $S = sH^T$ of the vector s (distorted word of length n) and the matrix H^T (H is the
  correcting-code check matrix with n - k columns and k rows, and H^T is the trans-
  pose of H).

‡ The error vector is customarily defined in the theory of correcting codes [106] as an
  n-dimensional vector in which zeros correspond to correct symbols and ones correspond to erroneous symbols of the code word.

tor describing the input states of the correcting block. The structure described by Rudnev and Khetagurov [123] for a corrector
correcting d = 1 error and the one described by Frantsis and Yanbykh [156] for the correction of d > 1 errors are realized on the
same principle.

Among the papers that deal with the synthesis of correcting
blocks we find two other papers, one by Frank and Yau [202] and
one by Levenshtein [66]. We have already discussed the former in
7.4. Its author consider the realization of encoders and decoders.
In the latter paper [66] the theoretical aspects of the synthesis of
decoders are investigated, and necessary and sufficient conditions
are determined for the existence of a decoding automaton when the
particular form of encoding is given.

7.7. Applications to Specific Devices. Few
papers are known in which specific d-reliable structures obtained
on the basis of effective correcting codes are described.† We can
cite six papers on the subject [287, 212, 174, 91, 160, 64], but only
some of the devices described in these references have actually
been put into practice.

In the first paper (in chronological order of publication)
[287] Russo investigates the reliability improvement of counters on
the basis of unit-error-correcting codes. The initial counter is
made up of triggers, NAND elements, and OR elements and has one
pulse input. The output information is extracted from the triggers.
The correcting code therefore encodes the states of the triggers
contained in the counter. The calculation of the number of parts
in the counters is programmed for a computer. Data are given on
the number of diodes required for the realization of a 1-reliable
counter with 5 to 8 states. It is remarked that the cost of the redundant counter is five or six times that of the nonredundant counter.

In [212] Görke investigates the structures of counters using
a Hamming code with different values of the coding distance.

An extension of this work is found in Beister and Görke's
paper [174]. They analyze various methods for increasing reliability
in application to decimal counters. Structures of 1-reliable decimal
counters are realized and compared in terms of several parameters.
In one version the states of the counter are assigned a Hamming

---

† Patents are not included in the present survey.

code (7.4). The counter is assembled from triggers and contains a failure-signaling block and two correctors (duplication). The complexity of the equipment is eight times the complexity of the nonredundant counter. The redundancy is diminished somewhat in the case of the diode net realization. Also realized in addition to these structures are decimal counters with triplication and subsequent majorization by the Moore — Shannon scheme, as well as counters using encoding but assembled from threshold elements. The latter structure is equivalent in complexity to the triplication structure. The least redundant structure is the one realized by the Moore — Shannon scheme, but the signalization of failure is difficult in this case. Of the five structures considered, preference was given to the realization of the majority structure, because it resulted in a relatively low redundancy and was readily amenable to failure signalization.

In [91] Nomokonov and Tolstyakov also investigate techniques for the improvement of counter reliability. A Berger code [106] is used to encode the states of the counter, because the authors found that the errors occurring in the counter correspond to asymmetric distortions of the code words used to code the states of the nonredundant counter.

In [160] Shekhovtsov investigates techniques for the synthesis of encoders that are d-reliable with respect to some of their elements.

Finally, in [64] Lapin develops a code for the correction of error batches and illustrates its application for the correction of errors on magnetic tape.

CHAPTER II

ASSURANCE OF STABILITY IN DISCRETE SYSTEMS

§1. Basic Concepts and Definitions

The demands for flawless operation on the part of discrete systems often cannot be satisfied merely by high reliability of their constituent elements and the instigation of measures to assure infallibility of the systems themselves; in their design, therefore, it is necessary to adopt measures in order to guarantee so-called functional "stability," or temporal reliability, through proper synthesis of the system structure, i.e., through a special choice of the number of elements in the system and of the character of their interconnections. Unstable operation of a discrete system can occur either due to the presence of random interference or due to interference brought on by a scatter of values of the time parameters† of the system elements. Whereas the influence of random interference or noise can be abated by proper filters or the assurance of d-reliability of the system, the elimination of interference caused by scatter of the time parameters of the elements can be taken care of in the appropriate stages of synthesis of the system by the introduction of structural redundancy, except that the introduction of structural redundancy in this instance differs from the introduction of structural redundancy for the assurance of d-reliability of the system. In this chapter, therefore, we interpret the stable operation of a system to mean execution of the functional algorithm by the discrete system when the time parameters of its elements are subject to variation. It is assumed in this connection that the variation of the time parameters of the elements does not exceed a stated value, for example that it is less than the time interval re-

---

† The "time parameter" notion embodies all factors associated with any change of state of the elements, including nonsimultaneous arrival of signals at the inputs, delay variations, etc.

quired for changes of the input states of the system. If the time
for a change of state of a particular element of the discrete system
is greater than the time between changes of input states, we say in
this case that a failure of the corresponding element has occurred
in the system. The assurance of discrete system stability has received attention in the literature since 1954 [223-225]. With the
subsequent progress of discrete systems (attended by growing
complexity, higher speeds, and the use of semiconductor elements)
the importance of stability has mounted steadily. More detailed
research on the problem has led to the formulation of various
stability types, the theoretical analysis of which has preoccupied a
number of authors, who have published more than a hundred papers.
More recently the stability assurance problem has been coordinated
with the assurance of infallibility (spatial reliability) and the quest
for simplicity in the structural realization of discrete systems.
In view of the specifics of the stability-assurance problem we shall
first present some fundamental concepts and definitions pertinent
to the problem.

Suppose that the discrete system is in the "stable" state,
i.e., that the output value of each element of its structure coincides
with the value of the Boolean function realized by the given element.
In the event that the output value of at least one element of the structure fails to coincide with the value of the Boolean function realized
by the given element the given state of the system is said to be
"unstable." Suppose that in transition from one state to another
several elements of the system simultaneously fall into an unstable
state. These elements may be either logic elements or memory
elements. The actual scatter of the time parameters of the elements or differences in the length of the networks for the transmission of signals produced by changes of states of the elements
could be responsible for nonsimultaneous changes of state of the
inputs of certain elements. Consequently, in the transition period
a change can take place in the prescribed order of changes of
state of the outputs of these and subsequent elements, whereupon
the output states of these and the subsequent elements can deviate
from the assigned states during the transition period. We say in
this case that in the given transition "race" conditions exist between the signals produced by the changes of state of the elements.
The following distinctions are made in the current literature between race conditions, depending on their origin.

○ Race conditions between input signals (between changes of state of the external inputs). We shall refer to this type of race henceforth as an I-race (short for "input race"). Accordingly, we shall refer to the stability of a system with respect to this type of race condition as I-stability.

○ Race conditions between input signals and memory element signals†; we shall call races of this type E-races and refer to the corresponding system stability condition as E-stability.

○ Race conditions between memory element signals‡ will be called M-races, and the corresponding system stability condition will be called M-stability.

○ Race conditions between logic elements signals,§ sometimes designated as races in the system logic generator, will be called L-races. The corresponding stability condition of the device will be called L-stability.

Race conditions can produce "transient" as well as "stationary" (or permanent) failures of the given functional algorithm of the system. In this case the race conditions are said to be "inadmissible."

It is important to realize that the influence of race conditions on the system operation depends on the information-processing mode in the system. Syncrhonous information processing makes it possible to circumvent the influence of I-, E-, and L-races and in some case M-races. Whenever it is required to eliminate race conditions in synchronous (clocked) systems, the same methods are used as in asynchronous systems. We shall therefore investigate the stability-assurance problem primarily in application to asynchronous systems. A classification of race conditions and techniques for the elimination of inadmissible race conditions is given in Fig. 16. Of the existing methods we first single out the elimination of races by means of band filters, which make it possible to remedy stray spikes or lengthen short

---

† These race conditions are sometimes called "essential hazards."
‡ These races are sometimes called "critical" races.
§ These race conditions are sometimes called "static hazards."

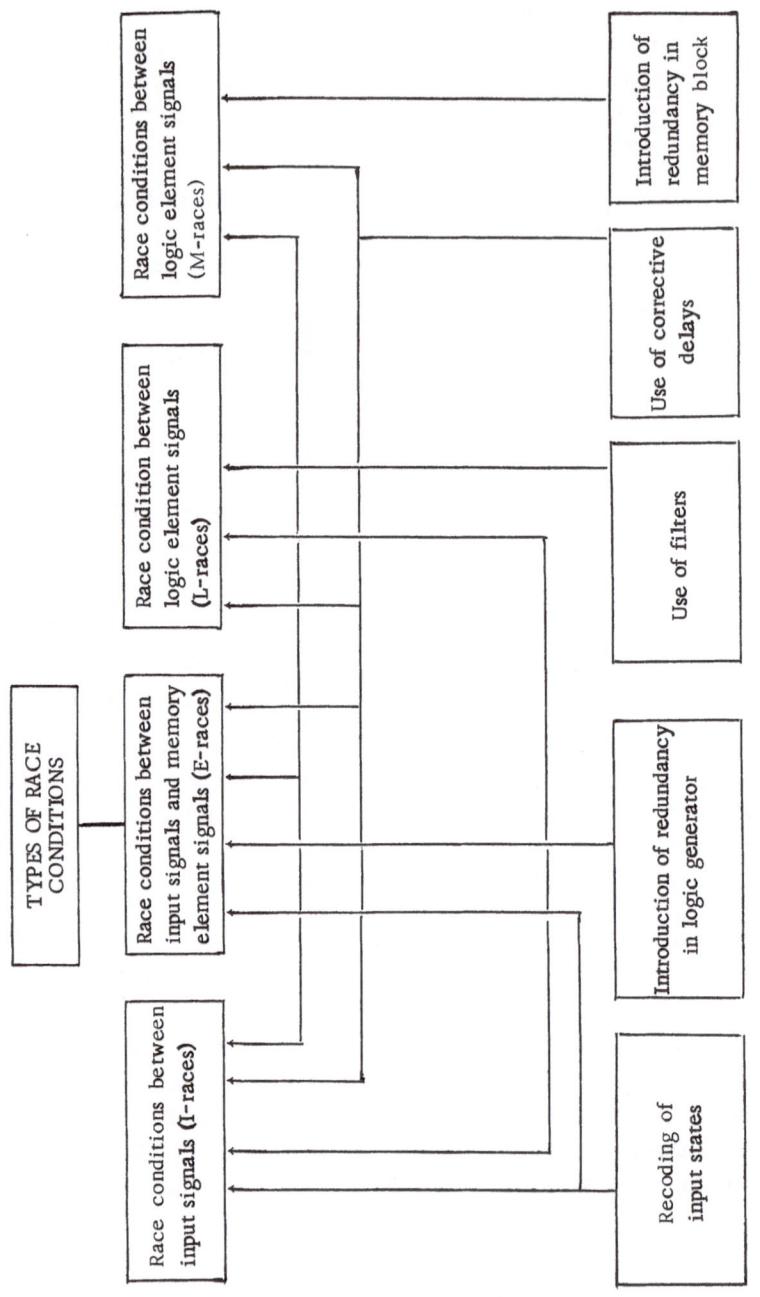

Fig. 16. Classification of race conditions in discrete systems.

spikes produced by races. The use of band filters affords a universal solution to the race-elimination problem, but it complicates the system and diminishes its speed, hence their use is limited. With regard to the analysis of other techniques for the elimination of inadmissible races it is convenient to introduce the concept of the n-dimensional unit-transition cube Q_n, which represents an n-dimensional cube in which each vertex is correlated with a code word of length n and any two vertices with adjacent code words (i.e., with Hamming distance equal to 1) are joined by an edge. In this case the state of the system, namely the state of the external inputs and outputs of the system elements (including output and internal elements) may be represented as a code word of length n, where the value of the i-th component corresponds to the value of the i-th variable corresponding either to a particular external input or to the output of a particular element of the structure.

We wish to examine a given transition from a state μ_f, denoting this transition as $\mu_f \rightarrow \mu_h$. Proceeding from the scatter of the time parameters of the elements that change their state in the given transition, we represent the transition as a path (or set of paths) in the cube Q_n (see Fig. 17).

We denote by $U_{f,h}$ the set of vertices of Q_n (including the initial and final) through which the given transitions can be made, i.e., we have for our example $U_{f,h} = \{000, 001, 010, 011\}$. Let us assume there exists another transition, namely the transition $\mu_q \rightarrow \mu_e$, $h \neq e$ (see Fig. 17). Let us assume that $U_{f,h} \cap U_{q,e} \neq \emptyset$, i.e., that there is at least one vertex of Q_n (in our example $U_{f,h} \cap U_{q,e} = \{000, 001, 010, 011\}$) through which both transitions can be made. If each transition can be terminated in a stable state different from the given one, depending on the time parameters of

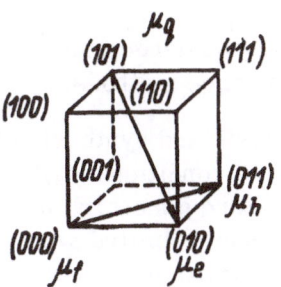

Fig. 17. Geometric interpretation of transitions in a discrete system.

the elements, we say that "inadmissible races," i.e., races producing stationary failure of the system functional algorithm are present in the discrete system. If each transition is terminated in the correct stable state, but output states different from the preceding and succeeding states occur in the transition process at the system external outputs corresponding to a certain part of the code word governing the state of the system, or an alternation of the preceding and succeeding output states occurs, we say that inadmissible races producing transient failure of the system functional algorithm are present in the system.

In order for stationary failures of the functional algorithm to be absent in a discrete system it is necessary and sufficient that for every pair of transitions to distinct stable states the intersection of sets of vertices of Q_n through which the given transitions are tenable be empty. Accordingly, it is necessary and sufficient for the absence of transient failures of the functional algorithm in a discrete system that the intersection of sets of transitions having distinct output states be empty. The so-called "decoupling" variable is introduced in order to guarantee the empty intersection of transitions. We compare each element output with its variable. Then the variable is called a decoupling variable for a pair of transitions if it retains a constant opposite value in those transitions, i.e., if its value in one transition is opposite the value in the other transition.

Consider the pair of transitions $\mu_f \rightarrow \mu_{f_1} \rightarrow \ldots \rightarrow \mu_h$ and $\mu_q \rightarrow \mu_{q_1} \rightarrow \ldots \rightarrow \mu_e$, where μ_h and μ_e are stable states, $h \neq e$. Let us assume that in some way or another sets of intermediate states $U_{f,h}$ and $U_{q,e}$ have been found through which both transitions can be made. Then in order for the condition $U_{f,h} \cap U_{q,e} = \emptyset$ to be satisfied it is sufficient that at least one decoupling variable exist for each pair of transitions $\mu_{\xi_i} \rightarrow \mu_{\xi_i + 1}, \mu_{\xi_j} \rightarrow \mu_{\xi_j + 1}$, $\mu_{\xi_i} \in U_{f,h}$, $\mu_{\xi_j} \in U_{q,e}$. For this reason the decoupling variable concept is widely used in stability-assurance methods.

In view of the fact that the analysis and synthesis of stable systems with regard for all types of race conditions still constitute an unsolved problem, the existing literature generally deals with the operational stability of a system in some limited aspect (as a rule, in application to one type of race condition). As we remarked above (see also Fig. 16), the assurance of system stability involves the introduction of a certain redundancy into the system, so that the isolated (stepwise) elimination of inadmissible races of each type can necessitate the introduction of a greater degree of redundancy than if all types of inadmissible races were eliminated concurrently. For this reason an effort is now being made (particularly in the United States) to develop a general theory of stable asynchronous control systems (see, e.g., Miller [260]).

§2. Elimination of Inadmissible I-Races

If the system working conditions call for the simultaneous change of state of several inputs of the system, then, as indicated above, I-races occur. One of the first papers to cope with this type of race condition was [223].

One of the main techniques used to eliminate I-races is the introduction of input constraints, i.e., a choice of input signals such that the value of only one input variable can change simultaneously. This is not always feasible, however, under the working conditions of the system. In this case the following stability-assurance techniques are used (see Fig. 16): a) the use of filters; b) delay correction at the inputs; c) recoding of the input states by the introduction of structural redundancy in the logic block.

The use of filters is a universal expedient. Inasmuch as they are no different in principle from the filters used to eliminate random noise, they have not been given too much attention in the literature for the elimination of functional interference. Our only reference is a paper by Yakubaitis [161], in which filters are investigated for the elimination of I-races. The use of corrective delay elements consists in the determination of the minimum number of additional delay elements and their positions of insertion into the structure of the system so as to reduce inadmissible races to admissible races or, more generally, to exclude races altogether. Referring once again to the cube Q_n, in which the vertices correspond to states of the system in the sense explicated above, the use of corrective delay elements makes it possible in a number of cases to ensure nonintersection of the sets $U_{f,h}$ and $U_{q,e}$ corresponding to the pair of transitions $\mu_f \rightarrow \mu_h$; $\mu_q \rightarrow \mu_e$ by a certain directionality of the transitions, i.e., by a choice of subsets $U'_{f,h} \subseteq U_{f,h}$ and $U'_{q,e} \subseteq U_{q,e}$ such as will guarantee nonintersection of the given transitions in Q_n. Corrective delay elements have found fairly wide application in the practical synthesis of discrete systems, although they cannot always be used to assure the required stability.

Given certain assumptions with regard to the relationship of the time parameters of the elements, I-stability can be assured by the methods described in [170], which are based on the introduction of structural redundancy in the system. In [170] Armstrong and others describe a method for the assurance of I-stability of a system by recoding of the input states (the method simultaneously assures E-stability). According to this method the structure realizing a given flow table A of the system (the flow table of a device is specified in the form proposed by Huffman [223]) consists of an encoder and, in series, a structure that realizes the flow table A' obtained from A by recoding of the input states and the introduction of an auxiliary state called a "spacer" state, † so that the transition from one input state in A' to another is effected only through the spacer state. This two-cycle operation is provided by an encoder containing a single delay element with time constant Δ such that after a change of a particular input variable the encoder generates a spacer state for a time Δ sufficient for the change of state of several inputs. The encoder consists of a logic block (converter) and triggers, the number of which is equal to the number of encoder outputs, i.e., the number of input states in table A. The spacer state has the code word 00, ..., 0; the i-th column in A' corresponds to the code word in which only the i-th component has a value of 1. With this realization of the system delay elements are not needed in the system feedback links.

The given method permits the I-stability assurance of a system independently of the relationships of the time parameters of the structural elements of the device. However, the auxiliary encoder in this case turns out to be rather complex.

§3. Elimination of Inadmissible E-Races

Suppose that the delay times of the memory elements are commensurate with the delay times of the logic elements (this can

---

† Any state relative to which all other input states of table A' are adjacent can be a "space" state.

happen, for example, in the case of a structure assembled from semiconductor elements). In this event E-races (essential hazards) are possible, which are essentially races such that the transmission delay of signals produced by variation of the values of the input variables before certain elements of the logic generator are equal to or greater than the transmission delays of signals produced by variation of the values of the internal variables. The possibility of improper operation of a system due to the presence of E-races was first brought to attention by Unger in [302]. He showed that the detection of E-races in a system requires the analysis of all transitions from the system flow table. E-races occur when it is possible to locate in the transition table a transition from an internal state s_i to some internal state s_j different from the one assigned and accessible by three successive changes of value of a particular input variable a_ε. If the transmission delay of the signal from a value change of that variable is greater than the transmission delay of the signal from a value change of the internal variables, the transition from s_i and to s_j is possible with one value change of a_ε, causing failure of the assigned functional algorithm. The techniques for the elimination of E-races are summarized in Fig. 16. The principal method for the assurance of E-stability is the introduction into each feedback link of a delay element, whose delay time is greater than the delay time of the logic elements. In [302], however, the need for the introduction of delay elements is systematically investigated, as a result of which it is shown that if E-, I-, M-, and L-races are absent in the system, the delay times in the feedback links of a discrete system can be as small as desired, i.e., they can be reduced to zero.

When E-races are present in the system, if the number of internal variables is equal to $2]\log_2 R[- 1$ (where R is the number of internal states) it is sufficient to have two more delays in a specially constructed block of the system memory. With the use of $2]\log_2 R[+ 1$ internal variables it is sufficient to have one more delay in the memory block, but the memory block turns out to be complex in either case. It is assumed in this connection that delays can occur either in the logic elements (element delays) or in the connecting wires (line delays). If line delays are neglected, as is justifiable in a number of situations, the class of systems in whose feedback networks the delay times can be made as small as desired and whose E-stability can still be assured is greatly expanded.

In [170], for example, Armstrong and others propose two methods for the assurance of E-stability without delay elements in the feedback networks, on the assumption that line delays are small and can be neglected. The first method has been described in detail in § 2. The second method is applicable when the system is I-, M-, or L-stable, the structure is synthesized from AND, OR, NOT elements with OR output elements, and the value of one input variable can change simultaneously at the system input. If E-races are present in this type of structure, the output of an OR element can, even with the variation of one input variable, in certain cases develop type $0 \to 1 \to 0$ or $1 \to 0 \to 1$ error spikes, which can lead to the stationary failure of the system functional algorithm.

We can investigate the causes of the onset of a $0 \to 1 \to 0$ spike in an example. In a certain transition with a change of value of the variable a_ε let the value of the internal variable y_i be required to remain zero, while the value of the variable y_j changes from 0 to 1. Let the expression for the excitation function of the i-th memory element contain a certain conjunction of the form $\bar{a}_\varepsilon \cdot y_j \cdot b_r$, which can have a valuation of 1 when y_j changes from 0 to 1 and when a_ε changes from 0 to 1, on the assumption that $\bar{a}_\varepsilon$ cannot instantaneously assume a value of 0. As a result, a $0 \to 1 \to 0$ spike appears at the output of the AND element realizing the given conjunction. The following transformation of the conjunction is proposed in the paper: $\bar{a}_\varepsilon \cdot y_j b_r = b_r \overline{\bar{a}_\varepsilon \cdot y_j} = b_r \cdot a_\varepsilon \vee \bar{y}_j$. In this case the path length for transmission of the value change of the variable y_j is greater than for a_ε. Next we investigate the causes of a $1 \to 0 \to 1$ spike at the output of an OR element of the structure realizing the excitation function of the i-th memory element. Let the value of one conjunction change fron 1 to 0 and the value of the other change from 0 to 1 simultaneously in the k-th transition (due to E-races). It is proposed in this case that the term $L_k \cdot \bar{z}_k$ be inserted in the expression for the excitation function of the i-th memory element, where L_k is a conjunction acquiring a value of 1 in the set of states through which the k-th transition can be made with simultaneous variation of the values of the input and internal variables, and z_k is a function acquiring at least a value of 1 in the set of k-th transition states in which the value of the excitation function for the i-th memory element is equal to zero.

An analogous approach to the assurance of E-stability has also been taken in [161].

§4. Elimination of Inadmissable M-Races

Huffman [223, 225] was the first to prove that if the simultaneous change of state of several memory elements is required in
a particular transition, the scatter of values of the time parameters
of those elements causes some to change their state more rapidly
than others and through the feedback network (in which the delay
time can be small) to possibly change the input states of certain
memory elements. As a result, the system can enter a stable state
different from the one prescribed, or an unassigned cyclic sequence
of states can develop. The possible techniques for the assurance of
M-stability of a discrete system are summarized in Fig. 16.

A classification (hierarchy) of trends toward the solution
of the problem of the synthesis of M-stable systems is given in
Fig. 18.

A great many papers have been devoted to the elimination
of M-races. They can all be grouped into one of two main problem
areas, depending on the auxillary problems that are solved concurrently, viz.: (1) minimization of the number of memory elements,
i.e., minimization of the length of the internal state codes; and (2)
structural simplification of the logic generator.

We now survey briefly the existing methods for solving each
of the indicated problems.

4.1. Methods for the Elimination of M-Races and Minimization of the Number of Memory Elements.†

It is convenient in the discussion of these methods
to represent the system functional algorithm in the form of a
transition graph $\Gamma = (S, V)$, where S is the set of vertices corresponding to the set of internal states and V is the set of edges corres-
ponding to the set of system transitions. The transition graph $I = (S, V)$ is said to be stable if the encoding of the internal states ensures the elimination of inadmissible M-races. The generation of
a stable transition graph can be initiated with verification of the
membership of the transition graph in the set of partial subgraphs ‡
of a unit-transition cube. If this occurs, we say that the graph $I = (S, V)$ is matched with the cube Q_n of dimension $n \geq]\log_2 R[$.

---

† A survey of work in this area may also be found in [146].

‡ Here and elsewhere we use the terminology of [5].

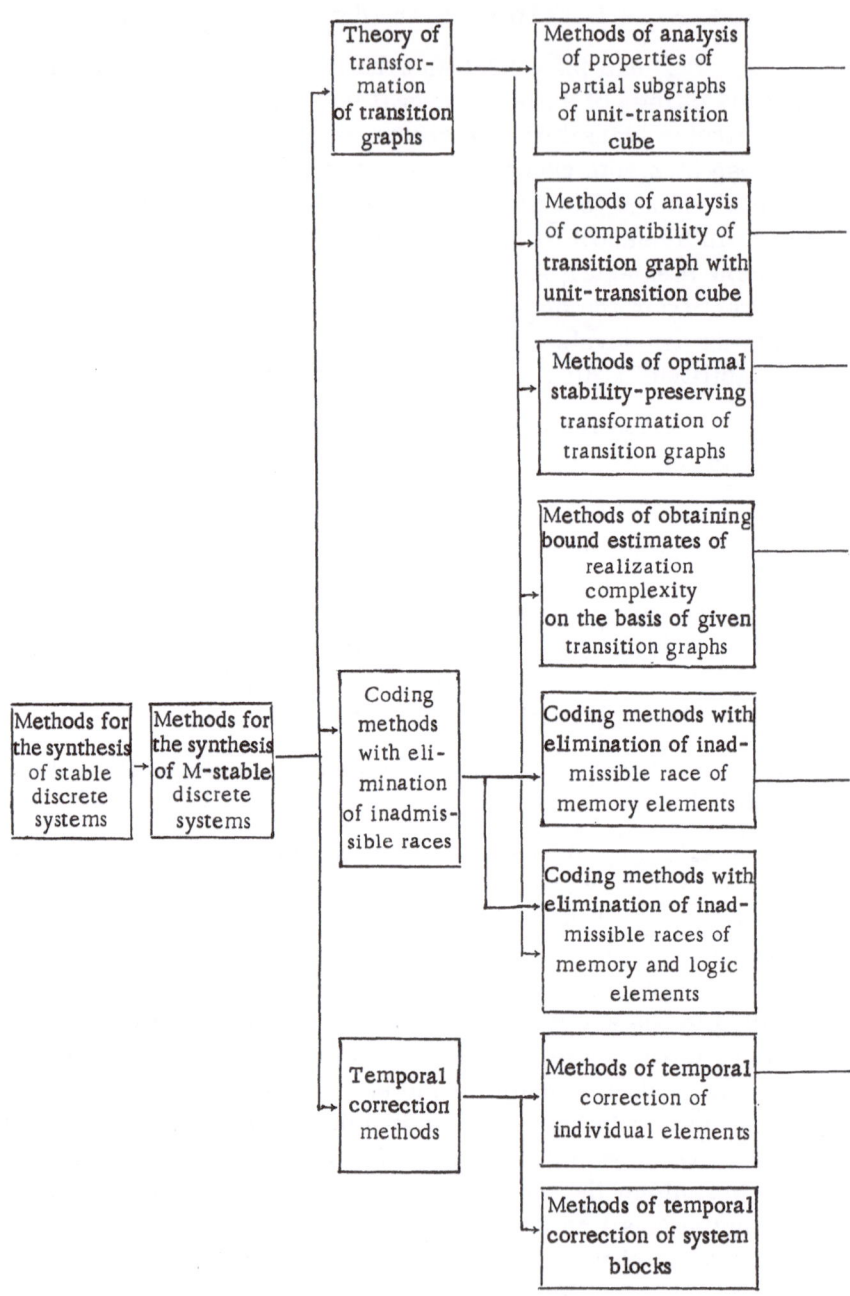

Fig. 18. Evolutionary classification (hierarchy) of methods

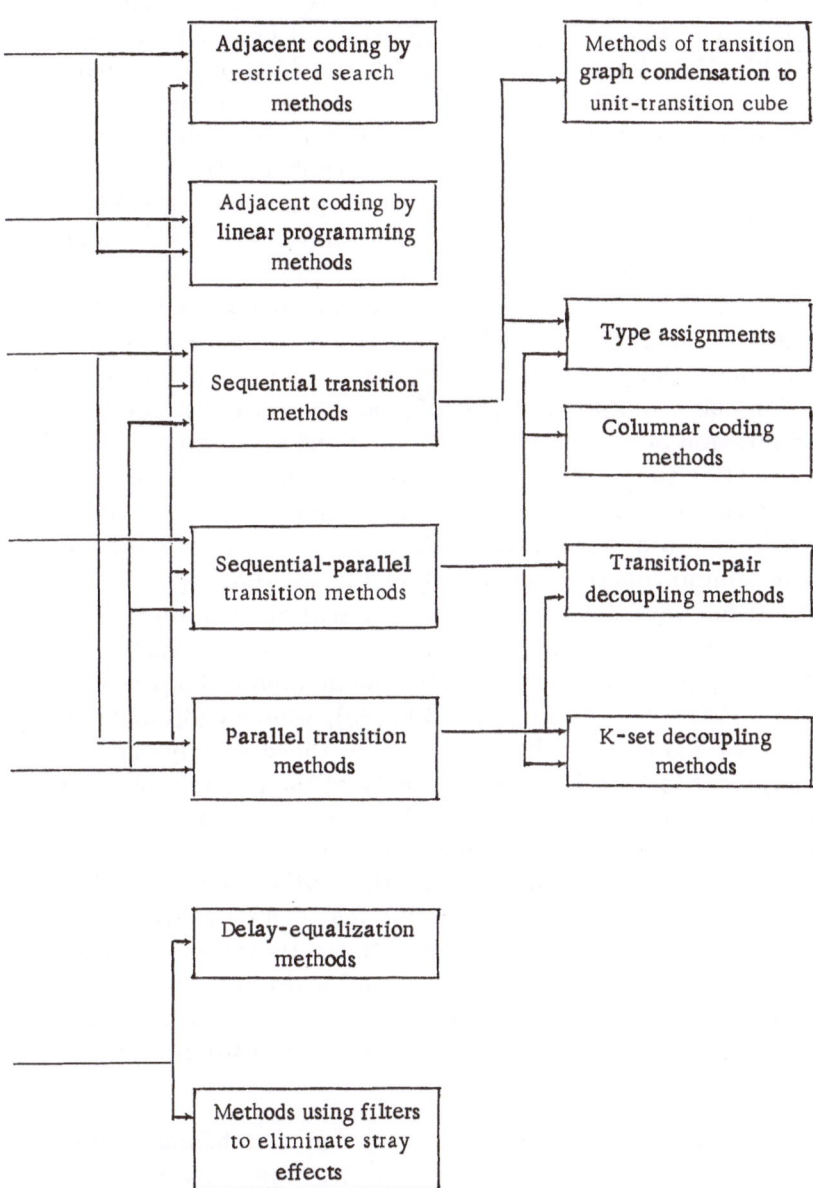

for the M-stability assurance of discrete systems.

4.1.1. Methods of Matching the Transition Graph with a Unit-Transition Cube.

It is readily seen that all possible M-races are eliminated by matching of the transition graph with a unit-transition cube, i.e., adjacent coding is a sufficient condition for M-stability of the system. It is clear that the compatibility problem for the transition graph and the unit-transition cube can be solved by inspecting all graph coding variants, the number of which is equal to $R!\, C_{2^n}^R$, where n is the code length.

We see at once that the number of variants grows so rapidly that, even on a digital computer, direct search becomes impracticable when $R > 10$. It is helpful in this respect to solve the problem on the basis of the properties of the partial subgraphs of the unit-transition cube. Kapitonova [47, 48] has shown, for example, that if the operation of "regular" partitioning and amalgamation is applied to a partial subgraph of the unit-transition cube, the resulting graph will also be a subgraph of a unit-transition cube of, in general, a different dimension. A partition of the set of vertices S of Γ into two nonintersecting subsets S_1 and S_2 ($S_1 \cup S_2$, $S_1 \cap S_2 = \emptyset$) is "regular" if for any vertex $s_i \in S_1(S_2)$ there at at most one vertex $s_j \in S_2(S_1)$ adjacent thereto in Γ. A graph $\overline{\Gamma}$ is called an amalgamation of Γ on this partition if there are isomorphisms $\varphi:\Gamma(S_1) - \overline{\Gamma}$ and $\psi:\Gamma(S_2) \to \overline{\Gamma}$ such that $\varphi(\alpha(S_1)] = \psi(\alpha(S_2)]$, where $\alpha(S_1)$ and $\alpha(S_2)$ are subsets of S_1 and S_2 such that if $s_i \in \alpha(S_1)$ there is only one $s_j \in \alpha(S_2)$ adjacent to it in Γ. This property of the partial subgraphs of the unit-transition cube was used as the basis of an algorithm for the matching analysis of a transition graph with a cube Q_n. The operations of regular partitioning and amalgamation are alternately applied to I until the resulting graph is one that can be easily coded by an adjacent code. In general the algorithm calls for sweeping-out of the entire tree of solutions. The code length turns out to be minimal. The technique is passive, i.e., if the graph does not meet the stated conditions, there are no recommendations for its transformation.

It has been shown by Zakrevskii [38] that a bichromatic distributed graph is a partial subgraph of Q_n. A graph is said to be "distributed" if every pair of nodes (i.e., vertices of degree higher than 2) is joined by a unique "pendant" chain of edges of length 2 or more (a "pendant" chain of edges is one that contains vertices of degree 2 that are not end-vertices). An algorithm is given in the cited reference for the transformation of a nonbichromatic distributed graph into a bichromatic by the introduction of auxiliary vertices. In this method, however, the code length turns out to be larger than or at best equal to the number of nodes of the graph, i.e., rather large; also, distributed graphs are rarely encountered among the transition graphs of systems.

In [52, 53] an algorithm is proposed for the adjacent coding of a transition graph; the initial code length is chosen as small as possible and increased until the graph has been encoded by an adjacent code (if possible). The algorithm can be used to decide the compatibility of a transition graph with a unit-transition cube.

In [221] Hlavička also describes an algorithm for the adjacent coding of a transition graph; the algorithm entails the sweeping-out of chains of the graph. The edges of the graph in this case are assigned the index numbers of the variables that must change their state in transition from one vertex to another. Recommendations are given for reducing the number of variables required. The number of variables depends in general on the choice of coding variables for the remaining edges.

4.1.2. Transition Graph Transformation Methods. For the case in which the transition graph cannot be matched with a unit-transition cube several authors have proposed methods for the transformation of the graph so as to reduce it to a form suitable for matching with the unit-transition cube. One of the necessary conditions for transformation of the graph is that the system corresponding to the transformed graph execute the same functional algorithm (correct to speed) as the primary device. We note that the graph transformation procedure can be carried out without determining the compatibility of the original graphs with Q_n by the methods of the preceding section. The number of auxiliary vertices required for transformation of the transition graph into a partial subgraph of the unit-transition cube is determined, on the one hand, by the efficiency of the method and, on the other, by the speed demands. Depending on the required speed, the transformation of the graph can be aimed at acquiring a system with maximum speed (by the use of parallel transitions) or at acquiring a system with less than maximum speed (by the use of sequential-parallel transitions). A transition is said to be parallel if the sequence of state changes of the memory elements is not fixed. The transition time from one state to another in this case is determined by the time required for a change of state of the "slowest" memory element.

If the sequence of state changes of certain memory elements is fixed in a transition, we have a sequential-parallel transition, which can become purely sequential if the sequence of stage changes of every memory element is fixed. When sequential or sequential-parallel transitions are used, a nonmaximum-speed system is obtained.

We first consider methods for the assurance of M-stability by the use of sequential transitions. The number of memory elements obtained by these methods is between the limits $]\log_2 R[$ and $2]\log_2 R[- 1$, although the upper bound of certain methods described below is greater than $2]\log_2 R[- 1$; however, the availability of "type assignments" † [223] makes it possible to guarantee the M-stability of a system using $2]\log_2 R[- 1$ memory elements.

We are interested in methods relating to the problem of "condensation" of the transition graph onto a unit-transition cube by the introduction of auxiliary vertices and edges in order to obtain a partial graph of the unit-transition cube. These methods rely solely on sequential transitions. A universal shortcoming of condensation methods is the rapid growth in the volume of computations required or the acquisition of solutions far from drastically nonminimal solutions. The condensation algorithm of Tomfel'd [147] comprises the following sequential operations: (1) minimization of the number of edges of the transition graph that do not coincide with edges of the cube; (2) transformation of noncoinciding edges by the introduction of auxiliary vertices; (3) "decoupling" of vertices that are incident to noncoinciding edges and cannot be transformed in step 2.

The advantage of this method is the fact that the number of memory elements turns out in many cases to be smaller than in other methods. In general, however, the procedure requires a direct sequential search of all matching variants, the number of which, as noted above, grows rapidly with R. The number of memory elements in this method does not exceed $2.58]\log_2 R[- 1$.

It may be necessary to precede the "condensation" algorithm by an algorithm for minimizing the number of "variant" transitions

---

† A "type assignment" refers to the encoding of internal states such as to eliminate inadmissible races of memory elements in the presence of all transitions between internal states in the primary flow table.

in the transition graph Γ [100], but the dimensions of the latter (i.e., the dimensions of the flow table) are few, even for completely defined flow tables.

Saucier [139] has described two programmed algorithms utilizing graph-theoretic methods. Not all the transitions occur in one cycle in the proposed coding scheme. In the first method the graph is constructed from edges providing the maximum number of transitions. After the introduction of each edge the graph is encoded with an adjacent code, where the following criteria are used for the transformation of edges that do not coincide with edges of the cube Q_n: minimum cube Q_n; minimum transition times; minimum number of auxiliary vertices. In the second method the first step is the same (construction of a graph that still admits adjacent coding), but in the ensuing stages a direct sequential search of all possible variants of the construction of such graphs is executed, and the realization of all transitions is guaranteed.

In a study by Gavrilov [11] the encoding of states is based on the delineation of a subgraph with predetermined parallel transitions and the "accommodation" of all other transitions to that subgraph (accomodation method).

Huffman [223] has proposed type assignments to assure M-stability. The number of memory elements in this method is large, being equal to $2]\log_2 R[- 1$ (for $0.75 \cdot 2^{]\log_2 R[} < R \le 2^{]\log_2 R[}$) and to $2]\log_2 R[- 2$ (for $2^{]\log_2 R[-1} < R \le 0.75 \cdot 2^{]\log_2 R[}$). Type assignments for $R = 8$ are shown in Fig. 19. Other type assignments are also proposed which require $2]\log_2 R[+ 1$ memory elements and yield higher speeds (shorter transition times from one state to another) than the first type assignments.

$y_3 y_4 y_5$	$y_1 y_2$ 00	01	11	10
000	0	1	2	2
001	0	1	3	3
011	0	0	2	3
010	1	1	2	3
110	4	5	4	4
111	4	5	5	5
101	6	6	6	7
100	7	7	6	7

Fig. 19. Type assignment (encoding) of eight internal states for M-stability assurance.

Ostrovskii [98, 99] has explored the possibility of diminishing the required number of memory elements below the type assignment level. He attempts to take account of the properties of the given transition table. The method entails the selection from the given transition graph of the largest subgraph that can be matched with the unit-transition cube of dimension $]\log_2 R[$ with the subsequent decoupling of connected transition pairs by the introduction of auxiliary variables. The number of memory elements required in this method is between the limits $]\log_2 R[$ and $3]\log_2 R[- 1$. However, the algorithm is rather difficult to manage, especially in the matching phase.

Tomfel'd [145] has proposed a method for condensing the transitions of each column of the transition table onto a unit-transition cube. The number of memory elements depends on the order in which the columns are chosen and on the order of "condensation" of transitions onto the cube.

Hazeltine [220] has proposed a technique for the encoding of states by columns of the flow table. The encoding rationale is as follows. Internal states are arbitrarily assigned a code of length $]\log_2 R[$ (it is stated in [220], of course, that the encoding should preferably be such as to assign adjacent code words to adjacent states, but there is no indication as to how this is to be accomplished). Suppose that the states $s_f = (110)$ and $s_h = (001)$. To the left of each code word we attach an auxiliary variable with zero valuation, i.e., state s_f corresponds to the code word 0110, and s_h to 0001. Then the given transition may be represented as follows: $0110 \rightarrow 1110 \rightarrow 1111 \rightarrow 1011 \rightarrow 1001 \rightarrow 0001$. The sequence of state changes (underscored) can be different. It is stated in the paper that it is always possible to organize the transitions in such a way that the required number of memory elements is not greater than for type assignments. However, the procedure by which the transitions are organized becomes rather complicated for a large number of states.

In [288] Saucier considers the possibility of reducing the estimation of type assignments. It is shown that a complete transition graph with $R \leq 12$ vertices can be so transformed as to make the required number of memory elements smaller than for type assignments (at best a decrease of one memory element was obtained). The idea is that the row set (set of equivalent states) corresponding to a particular row of the primary flow table is identical for every internal state (in terms of the number of rows). In summing up this group of methods, it is essential to point out the considerable difficulty and, as a result, the low dimensions of the problems that can be solved in practice.

Next we consider methods based on parallel transitions. The number of memory elements required by these methods fluctuates, depending on the specific flow table, from $]\log_2 R[$ to $2]\log_2 R[- 1$. We shall survey the relevant papers in their order of publication.

In [224] Huffman proposes type assignments of internal states. The basic notion of these assignments is that every row of the primary transition table A is replaced by a set of rows (equivalent states). This set of rows is called the row set. The author proposes a systematic procedure for encoding of the row sets, as a result of which each row set contains a row whose code word is adjacent to the code word of a particular row of any other row set. This "condensation" procedure requires at most $2]\log_2 R[- 1$ memory elements.

In [243] Liu investigates a method for the encoding of states by columns of the flow table. Let a certain column of the flow table contain k stable states. We compare them with different code words of length $]\log_2 k[$. We compare each unstable state with the code word of the state into which it is to be transferred. This operation is ensured by the presence of at least one decoupling variable for each transition pair for one input state. Then the code word of the internal state corresponding to the i-th row is obtained as the set of code words for all columns of the same row. Recommendations are given for reducing the number of memory elements and length of the resulting code. It is proved that when the code length is greater than the number of internal states it is possible to use equidistant codes of length $2]\log_2 R[- 1$.

In [133] Sagalovich presents an algorithm for obtaining the absolute minimum number of memory elements for Liu's encoding method. The number of memory elements in this case is assumed to be dependent on the chosen encoding variant. The method can be put in algorithmic form, but is only applicable to small flow tables.

Matsevityi and Denisenko [79] have proposed a method for the decoupling of "connected" transition pairs, i.e., pairs that do not have a decoupling variable.

In [299] Tracy investigates a method for minimizing the number of memory elements in application of the decoupling of connected transition pairs. For better understanding we introduce certain definitions from [41]. Every transition in the flow table is represented as a partial double modular partition, i.e., a row of length R (where R is the number of internal states) in which, if the transition pair $(\rho_a, s_i \rightarrow s_j)$, $(\rho_a, s_f \rightarrow s_h)$ is considered, a 1 in the ij position and a 0 in the fh position, or vice versa. All other positions of the partition are filled with a dash (−). These partitions can be connected by an implication relation, for example, the partition 100−0 implies 1−0− −0, i.e., the first can be derived from the second. The concept of compatibility of partitions and the concept of the common implicant of partitions are introduced: (101− −01) and (101 0−0−) are compatible, and (1010−01) is the common implicant. A set of partitions for which a common implicant can be found is said to be compatible. The common implicants of maximally compatible sets are called principal implicants.

The method calls for the formation of the matrix Q (matrix of all double modular partitions) and the subsequent determination of the set of principal implicants; then the subset of principal implicants that implies Q is also found. According to the data of A. E. Yankovskaya, the proposed method is relatively inefficient. Thus, the number of rows of Q (for an intermediate general-purpose computer) $\sigma(Q) < 15$, and the number of principal implicants is not greater than 10.

The ideas formulated in [79, 243, 299] are developed the furthest in papers by Zakrevskii and Yankovskaya [40, 41]. The elaboration of these techniques with the application of methods for the reduction of sequential search has made it possible to encode flow tables with dimensions up to R × M = 32,000.

Synthesis based on delay subnetworks is investigated in [161-163]. The internal states are not encoded after minimization of their number, but directly from the graph of stable states, in a manner analogous to the encoding procedure of Liu [243]. The length of the code word corresponding to each stable state in the j-th column of the flow table, however, is equal to the number of stable states in that column. The code word corresponding to the i-th stable state in the j-th column of the flow table contains only a single 1 in the i-th position. The code words of the internal

states are also formed as in Liu's method. After encoding of the
internal states the code length is minimized so that at least one
decoupling variable will occur for each transition pair associated
with one input state.

4.1.3. Temporal Correction Methods. These
methods assure the M-stability of a system for any internal-state
encoding variant.

In Caldwell's book [56] M-stability is achieved by the use
of a double modulator memory, one module of which preserves the
current internal state by maintaining its memory elements in the
stable state. The next internal state is sent to the other module,
whose output is not connected to the logic generator, so that the
races of its memory elements are admissible. Once the elements
of the second module have acquired a stable state, as signaled by
a special device, the outputs of the second module switch over to
the input of the logic generator. This causes the memory elements
of the second module to be held in the stable state, while the outputs of the first module become disconnected from the input of the
logic generator. This alternating operation of the memory modules
reduces all races to inadmissible races. However, the number of
memory elements required is $2]\log_2 R[+ 2$, and the switching of
the modules (using delay elements) must be fast enough, or the
time constants of the memory elements must be large enough to
preserve their state throughout the switching period of the modules.
The transition from one internal state to another requires two
cycles.

Sapozhnikov and Sapozhnikov [135] have used a similar
approach, i.e., arbitrary encoding of the internal states. The
method is based on the following concept. Each column of the flow
table of the system is partitioned into a series of columns by the
introduction of internal variables from auxiliary memory elements
so that each column of the new flow table contains just one stable
state, i.e., so that the races of the primary memory elements are
stable in the column. The transition from one internal state to
another with a change of input state is realized as follows: First
the state of the auxiliary memory elements changes (admissible
races), then a special memory element is energized, whereupon
the primary elements suffer a change of state and then the special
element is disengaged. The number of auxiliary memory elements

depends on the number of stable states in the flow-table column
and lies between the limits 2 and]$\log_2 R$[. The time required is
four cycles.

4.2. Methods of Obtaining M-Stable Systems with Minimum Logic Generator Complexity.

It is customarily assumed in the minimization of the number of memory
elements that a reduction in the number of excitation functions will
lead to simplification of the structure of the logic generator, but
there are several cases in which such an assumption is not justified.
To obtain the simplest logic generator structure with a minimum
(or nearly minimum) number of memory elements is one of the
complex problems involved in the synthesis of optimum systems in
the large. Before considering methods for structural simplification
of the logic generator in connection with the assurance of system
M-stability, we need to discuss the principal methods aimed ex-
clusively at simplifying the structure of the logic generator. At
the present time there are about a hundred papers on the encoding
of internal states for the purpose of imparting a simple structure
to the logic generator. The following main trends are discerned:
(1) consideration of adjacency on the unit-transition cube (fundamental work by Armstrong [168, 169]); (2) consideration of the
properties of partitions of the set of internal states in order to
reduce functional dependence (fundamental work by Hartmanis and
Stearns [217-219, 294]); (3) consideration of adjacency of encoding
columns (fundamental work by Dolotta and McCluskey [192]).

The basic notion in the first group is to find an encoding of
the vertices of the transition graph which will minimize the number
of edges that do not coincide with edges of the unit-transition cube;
this procedure, however, does not always produce a simpler structure on the part of the logic generator. Nevertheless, the given
approach has merit in that it can serve as the basis of a method
for obtaining an M-stable system, as in a study by Tomfel'd [147].

The largest number of papers on simplification of the logic
generator of a system fall in the second group. Utilization of the
partition properties of the set of internal states makes it possible
to reduce the functional dependence of the memory element excitation functions. However, not every system flow table contains
the needed partitions. Still, partition theory affords a delicate
tool for the analysis of the properties of specific system flow tables.

The consolidation of the first two approaches has resulted
in a rather effective method [192], in which adjacencies as well
as partition properties are taken into account.

Next we consider in brief the work done on the coordinated
solution of both stated problems (stability assurance and structural
simplification of the logic generator).

In [103–105], on the basis of the resulting partitions, Piil'
analyzes the complex problem of M-stability assurance of a system and diminution of the functional dependence of the memory
element excitation functions. In [103–104] he defines a compatible
partition π_c with normal partition π_n, where π_n is a partition with
the substitution property. A partition π_n having the substitution
property is a partition of the set of internal states into nonintersecting subsets such that a transition is made from any two states
of one subset for any input state to states of the same subset (including the original).

The partition π_c is compatible with π_n if: 1) $\pi_c \pi_n = \pi_0$ (the
null partition, i.e., each subset contains one and only one state); 2)
there exists a transition from states s_i and s_j of one subset of
π_c for input state ρ to states s_f and $s_h \neq s_h$, in which case states
s_f and s_h must belong to different subsets of π_n. It is shown that
if π_c and π_n exist for a system with R = 4 and if the value of the
variable y_1 is distributed in accordance with π_n and that of y_2 in
accordance with π_c, we then have, besides M-stability, indepen-
dence of the excitation function of the element Y_1 from the variable
y_2. It is shown that if the system can be decomposed into k (R $\leq 2^k$)
devices connected in series, each of which contains two states, then
the M-stability of the system can be assured and the functional
dependence of certain excitation functions reduced.

In [105] the "partition pair" concept is invoked to effect a
major expansion of the classes of flow tables that can be encoded
in such a way as to assure both M-stability and a reduction in
functional dependence. A common drawback is the complexity
of generating the partition pairs for a large number of states.

It is essential in concluding this section to point out that
the combined solution of both problems (stability assurance and
structural simplification of the logic generator) makes it possible
to optimize systems in the large. On the other hand, the combined

solution of both problems still entails fairly complicated and cumbersome synthesis algorithms.

§5. Elimination of Inadmissible L-Races

It can happen as a consequence of the different durations of the transition processes that the forward and inverse outputs of the same element can in a certain period of time develop like signals (zero or one). The event can produce an incorrect signal either at the inputs of the memory elements, resulting in stationary failure of the system functional algorithm, or at the external outputs of the system, resulting in transient variations of the functional algorithm. This type of situation can occur in a system otherwise invested with I-, E-, and M-stability.

Race conditions of this type (logic hazards) were first investigated by Huffman in [225], in which he proposed a method for their elimination by locating, in every pair of conjunctions in the excitation functions or output functions such that a certain variable a_i enters into one conjunction as a_i and into the other as a_i, a conjunction that preserves a value of 1 in transition from a state in which one of the conjunctions goes to 1 to a state in which the other conjunction goes to 1, where that transition is induced by a change of value of a_i. Suppose, for example, that we have the Boolean function $\varphi = a_i \psi \vee \overline{a_i} \varkappa$, which is realized by an AND–OR structure, so that a change of value of α causes an error spike of the form $1 \rightarrow 0 \rightarrow 1$, whereupon it is required to add to the function φ the conjunction $\psi \varkappa$, i.e., $\varphi = a_i \varphi \vee a_i \varkappa \vee \psi \varkappa$. It is shown in [18] that a function containing all prime implicants does not contain L-races. In [8] Vorzheva investigates the minimization of Boolean functions with regard for L-races.

Included in this category of race conditions are so-called dynamic hazards of the type $1 \rightarrow 0 \rightarrow 1 \rightarrow 0$ or $0 \rightarrow 1 \rightarrow 0 \rightarrow 1$. These race conditions in general present a lesser danger with respect to correct system operation. In [76] McCluskey presents an algorithm for the detection and elimination of both static and dynamic hazards.

§ 6. Stability Analysis of Discrete Systems

The stability of each of the foregoing types of race conditions can be determined from the flow table of the system. In most cases, however, the representation of the system functional algorithm in flow-table form is laborious and almost defies analysis. A number of investigations have been conducted in this connection to determine the presence of a particular type of inadmissible race conditions according to the specified excitation functions of the memory elements and outputs.

Methods have been proposed in [34, 115-122, 195] for the analysis of the functions with the use of ternary algebra in order to find one or more types of races. These methods have in common the disadvantage of analytical complexity in the case of problems involving large dimensions.

CHAPTER III

ASSURANCE OF OVERALL
(SPACE – TIME) RELIABILITY

The rapid increase in the complexity of systems and growing requirements for infallible and stable operation necessitate the
development of methods for the analysis and synthesis of stable
systems with regard for spatial reliability (infallibility) demands.

Only a few papers have been published to date dealing with
the solution of the problem of simultaneously assuring stability and
infallibility (temporal and spatial reliability). We shall therefore
consider only papers in which the assurance of M-stability and
d-reliability have been investigated. All the papers may be grouped into distinct categories, depending on what problems are to be
solved simultaneously. The following are representative of these
additional problems: (1) minimization of the number of memory
elements, i.e., minimization of the internal-state code length; and
(2) simplification of the logic generator structure. It is essential
to note that the solution of the latter problem is rather complex
and the only study we can cite in this direction is [63]. Its authors
investigate the problem of finding a code to assure stability and
spatial reliability with a concurrent reduction in the functional dependence of the memory element excitation functions.

We begin with a discussion of the first problem. As we
mentioned in Chapter I, Gavrilov [9] has proposed the encoding of
internal states with a code correcting d errors of the memory
elements. In general, however, this encoding tends to lower the
operational stability of the system. A number of authors have
attempted the simultaneous assurance of spatial reliability and
M-stability. In [129], for example, Sagalovich proposes a relatively simple method for the construction of a code to assure M-stabil-

ity and d-reliability by the formulation of type assignments similar
to [223]. Consequently, the required number n of memory elements
turns out to be relatively large, being equal to $(2d + 1) (N - 1) \leq n \leq (2d + 1) (N + 1) - 1$, where $N = 2k \geq R$. It is assumed that
memory element failures can occur both during transitions and
during periods in which the device is in a stable state. A method
is also proposed, consisting in the $(2d + 1)$-fold iteration of a code
that assures M-stability of the system.

In [86] Nemsadze obtains a lower estimate of the number of
memory elements for the simultaneous assurance of d-reliability
and M-stability of a system.

In [131] Sagalovich effects M-stability and d-reliability by
means of a columnar error-correcting code (his method is similar
to that of Liu for the assurance of M-stability). He also gives an
algorithm for minimization of the composite code.

In [133], after maximum reduction of the code length (for
the assurance of M-stability), Sagalovich adopts the code symbols
for each column as information symbols and invokes conventional
linear encoding to form check symbols for each column, whereupon the code length is minimized.

A similar problem is solved in [95-97, 111], but it is assumed that memory element failures can only occur when the system is
in a stable state. This assumption permits certain transitions to be
realized as parallel-sequential transitions, so that the required
number of memory elements is diminished. In [97], for example,
Ostianu and Potekhin propose type assignments requiring (for $d = 1$)
$3]\log_2 R[$ memory elements (the type assignments for $R = 8$ are
shown in Fig. 20). (Analogous assignments can be generated for
$d \geq 1$ as well.) The resulting type assignments are independent of
the properties of the specific flow table, hence a further reduction
in the required redundancy can be counted on if the specifics of the
particular flow table are taken into account. Ostinau and Potekhin
[96] have shown that if the transition graph of a system is encoded
with an adjacent code of length m, it will suffice in the case $d = 1$
to double the given code and add one auxiliary internal variable so
that adjacent vertices of the transition graph will be assigned different values of that auxiliary variable. This approach affords a
simple technique for the synthesis of a code having the necessary
properties, and the code length $n = 2m + 1$, where $m \geq]\log_2 R[$.

Type assignment (encoding) table. The four header rows give the column codes $y_3\,y_4\,y_5\,y_9$ (read top-to-bottom for each column); the left column gives the state $y_8\,y_7\,y_6\,y_2\,y_1$. Circled values mark the centers of the spheres.

$y_8\,y_7\,y_6\,y_2\,y_1$ \ $y_3\,y_4\,y_5\,y_9$	0000	1000	0100	1100	0010	1010	0110	1110	0001	1001	0101	1101	0011	1011	0111	1111
0 0 0 0 0	Ⓞ	0	0	1	0	0	2	2	0	0	0	0	2	2	2	2
0 0 0 0 1	0	1	0	1	1	1	2	2	0	0	0	1	2	2	2	2
0 0 0 1 0	0	1	0	1	0	0	3	3	1	1	0	1	3	3	3	3
0 0 0 1 1	1	①	0	1	1	1	3	3	1	1	1	1	3	3	3	3
0 0 1 0 0	0	0	2	2	2	3	②	2	0	0	0	0	2	2	2	2
0 0 1 0 1	0	0	3	3	2	3	2	3	0	0	0	0	2	3	2	2
0 0 1 1 0	1	1	2	2	2	3	2	3	1	1	1	1	2	3	3	3
0 0 1 1 1	1	1	3	3	2	3	3	③	1	1	1	1	3	3	3	3
0 1 0 0 0	0	1	0	1	0	0	0	1	4	4	6	6	4	4	6	6
0 1 0 0 1	0	1	0	1	0	1	0	1	5	5	7	7	5	5	6	7
0 1 0 1 0	0	1	0	1	0	1	0	1	4	4	6	6	4	4	6	7
0 1 0 1 1	0	1	0	1	1	1	0	1	5	5	7	7	5	5	7	7
0 1 1 0 0	2	3	2	2	2	3	2	3	4	4	6	6	4	4	6	6
0 1 1 0 1	2	3	2	3	2	3	2	3	4	5	7	7	5	5	7	7
0 1 1 1 0	2	3	2	3	2	3	2	3	4	5	6	6	4	4	6	6
0 1 1 1 1	2	3	3	3	2	3	2	3	5	5	7	7	5	5	7	7
1 0 0 0 0	0	0	2	2	0	0	2	2	4	5	4	5	4	4	4	5
1 0 0 0 1	1	1	3	3	1	1	2	3	4	5	4	5	4	5	4	5
1 0 0 1 0	0	0	2	2	0	0	2	3	4	5	4	5	4	5	4	5
1 0 0 1 1	1	1	3	3	1	1	3	3	4	5	4	5	5	5	4	5
1 0 1 0 0	0	0	2	2	0	0	2	2	6	7	6	6	6	7	6	7
1 0 1 0 1	0	1	3	3	1	1	3	3	6	7	6	7	6	7	6	7
1 0 1 1 0	0	1	2	2	0	0	2	2	6	7	6	7	6	7	6	7
1 0 1 1 1	1	1	3	3	1	1	3	3	6	7	7	7	6	7	6	7
1 1 0 0 0	4	4	4	4	6	6	6	6	④	4	4	5	4	4	6	6
1 1 0 0 1	4	4	4	5	6	6	6	6	4	5	4	5	5	5	6	6
1 1 0 1 0	5	5	4	5	7	7	7	7	4	5	4	5	4	4	7	7
1 1 0 1 1	5	5	5	5	7	7	7	7	5	⑤	4	5	5	5	7	7
1 1 1 0 0	4	4	4	4	6	6	6	6	4	4	6	6	6	7	⑥	6
1 1 1 0 1	4	4	4	4	6	7	6	6	4	4	7	7	6	7	6	7
1 1 1 1 0	5	5	5	5	6	7	7	7	5	5	6	6	6	7	6	7
1 1 1 1 1	5	5	5	5	7	7	7	7	5	5	7	7	6	7	7	⑦

Fig. 20. Type assignment (encoding) of eight internal states for the assurance of M-stability and d-reliability of a discrete system (the centers of the spheres are circled).

The code synthesis technique for $d \geq 1$ is analogous, so that for $d \geq 1$ we have $n = (d + 1)m + d$. This code synthesis technique affords simultaneous d-reliability of the memory block and M-stability, although certain transitions will be sequential-parallel.

If triggers are used for the memory elements, as proposed in Ostianu and Potekhin's paper [97], the M-stability of a system

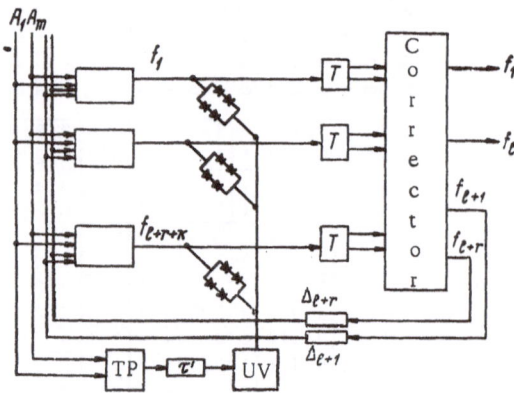

Fig. 21. Redundancy model for the assurance of
M-stability and d-reliability.

can be assured by cycling of the corrector, as illustrated in **Fig. 21.**
With a change in the input state the trigger pulse (TP) block generates a pulse to trigger the univibrator (UV), and after a time τ',
where $\tau' \geq \max(\tau_{fb})$ and τ_{fb} is the delay time of the functional
block, the UV generates a decision signal q of length τ'', where
$\tau'' \geq \max_i \tau_i'$, i = 1, 2, ..., $l + r + k$, and τ_i' is the firing time
of the i-th trigger. If min $(\Delta_j) > \tau''$, j = 1, 2, ..., $l + r + k_2$, where
min (Δ_j) is the minimum value of the delay time of the corrector
and functional block, inadmissible M-races will not occur.

LITERATURE CITED[†]

1. Artyukhov, V. L., and Blinder, M. I., An Economical composition technique based on logical triplication with interweaving, Izv. Leningrad. Élektrotekh. Inst., No. 68, pp. 189-193 (1968).

2. Akhmetkaziev, A., Encoding of the internal states of a finite automaton with antirace codes, in: Theory of Automata and Methods for the Formalized Synthesis of Computers and Systems (Seminar Proceedings, 1968) [in Russian], No. 7, Kiev (1969), pp. 18-49.

3.[*] Babaev, D. A., and Babaeva, D. G., An optimum composition problem, in: Analysis and Synthesis of Automatic Control Systems [in Russian], Nauka, Moscow (1968), pp. 120-123.

4.[*] Barlow, R. E., and Proschan, F., Mathematical Theory of Reliability [Russian translation], Sov. Radio, Moscow (1969), 488 pages (English edition: see [172]).

5.[*] Berge, C., Theory of Graphs and Its Applications, (Methuen) Barnes and Noble, New York (1962).

6. Blinder, M. I., and Lapshin, V. N., Investigation of the method of multichannel combinational composition of relay contact structures, Izv. Leningrad. Élektrotekh. Inst., No. 68, pp. 52-55 (1968).

7.[*] Braslavskii, D. A. Quorum elements for devices with functional redundancy, in: Variable-Structure Systems and Their Application in Flight Automation Problems [in Russian], Nauka, Moscow (1968), pp. 217-225.

8.[*] Vorzheva, V. V., Method of obtaining a race-free partial minimal form of a Boolean function, in: Abstract and Structural Theory of Relay Devices [in Russian], Nauka, Moscow (1966), pp. 112-116.

9.[*] Gavrilov, M. A., Structural redundancy and reliability of relay devices, Proc. First Internat. Congr. IFAC, 1960 [in Russian], Izd. AN SSSR, Moscow (1960), 25 pages.

10.[*] Gavrilov, M. A., Signalization and prognostication of failures in discrete control devices with structural redundancy, in: Discrete and Self-Adaptive Systems, Proc. Second Internat. Congr. IFAC, Basel [in Russian], Nauka, Moscow (1965), pp. 211-220.

11.[*] Gavrilov, M. A., Structural theory of relay devices, Part IV, Lectures at the All-Union Correspondence Institute of Electrical Engineering (VZÉI) [in Russian], Moscow (1964), 154 pages.

---

†References cited in the text of the book are marked with asterisks.

12.* Gavrilov, M. A., Techniques for the construction of reliable discrete (relay) devices, Seminar: Methods for the Construction of Reliable Systems from Unreliable Elements [in Russian] Kiev (1969), 66 pages.

13.* Gavrilov, M. A., Ostianu, V. M., Rodin, V. N., and Timofeev, B. L., Realization of discrete corrector networks, Dokl. Akad. Nauk SSSR, 123(6):1025-1028 (1958).

14.* Georgievskii, V. B., Synthesis of reliable systems from unreliable elements by the feedback method (I), Avtomatika, No. 1, pp. 49-57 (1965).

15.   Georgievskii, V. B., Synthesis of reliable systems from unreliable elements by the feedback method (II), Avtomatika, No. 3, pp. 41-45 (1965).

16.* Gindikin, S. G., and Muchnik, A. A., Solution of the completeness problem for systems of logic-algebraic functions with unreliable realization, in: Problems in Cybernetics, No. 15 [in Russian], Nauka, Moscow (1965), pp. 65-84.

17.* Glinski, G., Information-theoretic problems in the theory of unreliable automata [Russian translation], in: Theory of Finite and Probabilistic Automata, Nauka, Moscow (1965), pp. 280-300.

18.* Glushkov, V. M., Synthesis of Digital Automata [in Russian], Fizmatgiz, Moscow (1962), 476 pages.

19.* Gnedenko, B. V., Belyaev, Yu. K., and Kovalenko, I. N., Mathematical aspects of reliability theory, in: Probability Theory, Mathematical Statistics, and Theoretical Cybernetics, 1964 (Itogi Nauki, Ser. Matem.), VINITI AN SSSR, Moscow (1966), pp. 7-53.

20.* Gnedenko, B. V., Belyaev, Yu. K., and Solov'ev, A. D., Mathematical Methods in Reliability Theory [in Russian], Nauka, Moscow (1965), 524 pages.

21.* Gol'denberg, L. M., and Malev, V. A., Majority composition of a digital logic device, Radiotekhnika, 24(2):9-16 (1969).

22.* Gribanov, G. N., Synthesis of redundant single-cycle logic devices with majority decoding, Trudy Ural. Politekh. Inst., 168:119-124 (1968).

23.* Gridin, Yu. V., Maintaining the operability of a digital automaton in the event of malfunction of memory elements, in: Seminar: Theory of Automata [in Russian] No. 3, Kiev (1966), pp. 71-83.

24.* Gurvits, E. A., Synthesis of Polysynchronous Discrete Devices [in Russian], Svyaz', Moscow (1969), 174 pages.

25.   Denisenko, E. L., Some constraints imposed on the encoding of internal states of multicycle control devices, Kibernetika, No. 5, pp. 40-53 (1969).

26.* Diduk, N. N., Synthesis of Reliable Systems from Unreliable Logic Elements (Scientific Council on Cybernetics, Academy of Sciences of the USSR, Kiev House of Scientific-Technical Propaganda, Materials of Scientific Seminars on Theoretical and Applied Problems of Cybernetics, Seminar: Problems in the Mathematical Theory of Electronic Digital Machines) [in Russian], Kiev (1964), 32 pages.

27.* S. M. Domanitskii, Methods for the Synthesis of Redundant Structures with Restoring Organs (Scientific Council on Cybernetics, Academy of Sciences of the USSR, Kiev House of Scientific-Technical Propaganda, School-Seminar: Methods for the Synthesis of Reliable Systems from Insufficiently Reliable Elements) [in Russian], Kiev (1968), 35 pages.

28.* Domanitskii, S. M., Selection of the optimum adaptation algorithm for redundant structures with a variable-threshold restoring organ, Avtomat. i Telemekhan., No. 8, pp. 167-174 (1969).

29.* Domanitskii, S. M., and Prandgishvili, I. V., Reliable logic elements and output amplifiers with a redundant structure, Avtomat. i Telemekhan., 25(4) (1964).

30.* Doncheva, N., Encoding of the internal states of finite automata, Izv. Inst. Tekh. Kibernetika, 7: 61-77 (1967).

31.* Doncheva, N., Synthesis of a corrector in finite automata with structural redundancy, Izv. Inst. Tekh. Kibernetika, 8: 83-120 (1968).

32.* Druzhinin, G. V., Reliability of Automatic Devices [in Russian], Izd. Énergiya, Moscow-Leningrad (1964), 320 pages.

33.* Druzhinin, G. V., Reliability of Automatic Systems (revised and enlarged 2nd edition) [in Russian], Énergiya, (1967), 527 pages.

34.* Ershova, É. B., Roginskii, V. N., and Sutorikhin, N. B., Fundamentals of Relay Automation [in Russian], Svyaz', Moscow (1969), 176 pages.

35.* Zakrevskii, A. D., Method of synthesis of functionally-stable automata, Dokl. Akad. Nauk SSSR, 129(4):729-731 (1959).

36.* Zakrevskii, A. D., Functional stability of relay networks, Collected Works of SFTI on Automation and Computer Technology [in Russian] (1961), pp. 112-126.

37.* Zakrevskii, A. D., Functional stability of network states with respect to false closures, Avtomat. i Telemekhan., 25(9):1336-1343 (1964).

38.* Zakrevskii, A. D., Flow-table transformation algorithms eliminating race conditions, Trudy Sibirsk. Fiz.-Tekh. Inst. pri Tomsk. Univ., No. 47, pp. 48-55 (1965).

39.* Zakrevskii, A. D., On the synthesis of logic networks converting information into Ming code, in: Cybernetics in the Service of Communism [in Russian], Vol. 3, Énergiya, Moscow-Leningrad (1966), pp. 81-86.

40.* Zakrevskii, A. D., and Yankovskaya, A. E., Encoding of internal states of an automaton, Trudy SFTI, No. 51, Tomsk (1968).

41.* Zakrevskii, A. D., and Yankovskaya, A. E., Practical algorithms for the encoding of internal states of asynchronous automata, Avtomat. i Vychisl. Tekh. No. 3, pp. 15-21 (1969).

42.* Zarenin, Yu. G., Informational redundancy as a method for enhancing the reliability of arbitrary transformations of discrete information, School-Seminar: Methods for the synthesis of Reliable Systems from Insufficiently Reliable Elements [in Russian], Obshch. Znanie, Ukr.SSR, Kiev Dom Nauch.-Tekh. Propagandy, Kiev (1968), p. 32.

43.* Zarenin, Yu. G., and Savchenko, Yu. G., Nonlinear Codes correcting arbitrary error systems, Proc. Third Conf. Theory of Information Transmission and Encoding [in Russian], Moscow (1967).

44.* Zarovnyi, V. P., On self-adaptive decoding automata, Dokl. Akad. Nauk SSSR, 187(3):525-527 (1969).

45. Zenin, V. M., Use of constant-weight codes for the encoding of internal states of control automata, in: Theory of Accuracy and Reliability of

Cybernetic Systems, Seminar Proceedings [in Russian], No. 2, 1968, Kiev (1969), pp. 88-107.

46.* Ivas'kiv, Yu. L., and Ryakin, O. M., An information model of unreliable combinational networks, Kibernetika, No. 6, pp. 41-46 (1966).

47.* Kapitonova, Yu. V., Encoding of abstract automata (a special problem), KDNTP Seminar on the Theory of Automata [in Russian], No. 1, Kiev (1962).

48.* Kapitonova, Yu. V., Encoding of abstract automata with C-codes, Kibernetika, No. 1, pp. 40-44 (1965).

49.* Karpovskii, M. G., Synthesis of discrete devices with the detection and correction of errors of a given multiplicity, Izv. Leningrad. Élektrotekh. Inst., No. 56, Part 2 (1966).

50.* Karpovskii, M. G., Finite automata with error correction, in: Problems in the Theory of Electronic Digital Mathematical Machines, Seminar [in Russian], No. 3, Kiev (1967), pp. 83-99.

51.* Karpovskii, M. G., Finite automata with error correction and detection in: Computer Engineering and Cybernetic Problems [in Russian], No. 5, Leningrad. Univ., Leningrad (1968), pp. 60-73.

52.* Kartashev, V. I., Automation of the encoding of states of an automaton, in: Cybernetics and Computer Engineering [in Russian], Nauk. Dumka, Kiev (1964), pp. 45-51.

53.* Kartasheva, S. P., and Kartashev, V. I., An algorithm for the encoding of automata, in: Cybernetics and Computer Engineering [in Russian], Nauk, Dumka, Kiev (1964), pp. 32-44.

54. Kasparov, G. A., Application of the state concept in the synthesis of redundant structures, Trudy Leningrad Inst. Aviats. Priborostr., No. 63, pp. 125-134 (1969).

55.* Kirienko, G. I., Self-correcting networks of functional elements, in: Problems in Cybernetics [in Russian], No. 12, Nauka, Moscow (1964), pp. 29-37.

56.* Caldwell, S. H., Switching Circuits and Logical Design [Russian translation], Izd. In. Lit., Moscow (1962), 737 pages (English edition: see [206]).

57.* Khrishtal', V. Z., Method of synthesis of relay devices insensitive to element failures associated with unreliable input functions, Izv. AN SSSR, Tekh. Kibernetika, No. 4, pp. 122-126 (1966).

58.* Krishtal', V. Z., Application of batch-error-correcting codes to the synthesis of reliable relay devices, in: Reliability Synthesis of Digital Automata [in Russian], Kiev (1963), pp. 7-10.

59.* Krishtal', V. Z., Method of synthesis of relay devices insensitive to functional element failures, in: Problems in Technical Cybernetics [in Russian], Nauka, Moscow (1966), pp. 16-21.

60.* Krishtal', V. Z., Method of synthesis of relay devices insensitive to asymmetric failures, Avtomat. i Telemekhan., No. 5, p. 103 (1966).

61.* Krishtal', V. Z., and Ostianu, V. M., Use of redundancy to increase the reliability of relay devices, Proc. Third All-Union Conf. Automatic Control (Technical Cybernetics), 1965, Self-Adaptive Systems [in Russian], Nauka, Moscow (1967), pp. 356-362.

62.* Khristal', V. Z., and Ostianu, V. M., Increasing the reliability of relay devices composed of functional elements, Third Conf. Theory of Information Transmission and Encoding, Sec. III [in Russian], Moscow (1967), pp. 10-22.

63.[*] Lazarev, V. G., Piil', E. I., and Turuta, E. N., Reliability of discrete automata, Kibernetika, No. 1, pp. 1-7 (1968).

64.[*] Lapin, V. S., Automatic correction of batched errors on magnetic tape, Probl. Peredachi Inform., 4(1):28-34 (1968).

65.[*] Levenshtein, V. I., On self-adaptive automata, in: Theory of Finite and Probabilistic Automata [in Russian], Nauka, Moscow (1965), pp. 150-153.

66.[*] Levenshtein, V. I., Decoding automata invariant with respect to the initial state, in: Problems of Cybernetics [in Russian], Vol. 12, Nauka, Moscow (1964), pp. 125-136.

67.[*] Levin, V. I., Probabilistic Analysis of Unreliable Automata [in Russian], Zinatne, Riga (1969), 234 pages.

68.[*] Leont'ev, L. P., Determination of the optimum technique for the interconnection of primary and standby elements, in: Automation and Computer Technology [in Russian], No. 6 (1963), pp. 177-191.

69. Leont'ev, L. P., and Margulis, A. M., Possible ways of enhancing reliability by the use of series-parallel continuous-standby networks, in: Automation and Computer Engineering [in Russian], No. 12, Zinatne, Riga (1966), pp. 155-160.

70.[*] Litvinov, V. A., and Savchenko, Yu. T., Magnetic element with a branched magnetic circuit as a restoring organ for redundant automata, Izv. AN SSSR, Tekh. Kibernetika, No. 4, pp. 102-106 (1966).

71.[*] Löfgren, L., On the self-restoration concept, redundancy limits of networks, and carrying capacity of a computation channel [Russian translation], in: Theory of Finite and Probabilistic Automata, Nauka, Moscow (1965), pp. 345-370.

72.[*] Löfgren, L., Self-restoration as the limit for automatic error correction [Russian translation], in: Self-Organization Principles, Mir, Moscow (1966), pp. 226-283.

73.[*] Lysikov, V. T., and Mamedli, É. M., Selection of digital controller structure for a predetermined reliability, in: Problems in the Theory of Special-Purpose Computer Structures [in Russian], Sov. Radio, Moscow (1968), pp. 16-28.

74.[*] Lyakhovich, V. F., Certain methods for the synthesis of relay devices with input and output error correction, Materials of the Seminar on Cybernetics [in Russian], No. 3, Academy of Sciences of the USSR, Kishinev (1968), 32 pages.

75.[*] Madatyan, Kh. A., Synthesis of networks correcting contact openings, Dokl. Akad. Nauk SSSR, 159(2):290-293 (1964).

76.[*] McCluskey, E. J., Jr., Transients in combinational logic circuits, Redundancy Techniques for Computing Systems, Spartan Books, Washington, D. C. (1962), pp. 9-46.

77. Malyshev. V. A., Possibility of computing discrete functions with a certain probability, in: Discrete Analysis [in Russian], No. 5, Novosibirsk (1965), pp. 27-30.

78.[*] Malyugin, V. D., A method for the synthesis of redundant networks, in: Cybernetics [in Russian], Nauka, Moscow (1967), pp. 144-150.

79.[*] Matsevityi, L. V., and Denisenko, E. L., Encoding of internal states of certain multicycle devices, Kibernetika, No. 1, pp. 1-6 (1966).

80. Matsevityi, L. V., and Chaika, N. S., Encoding of the states of a multicycle device with allowance for certain logical properties of the elements used, Kibernetika, No. 4, pp. 25-32 (1966).

81. Miyata Masatika, Realization of reliable logic structures from unreliable elements, Trans. Inst. Electron. and Commun. Engrs. Japan, C52(6):362-363(1969).

82. Moisil, G. C., Algebraic Theory of Discrete Automatic Devices (Russian translation from Romanian), Izd. In. Lit., Moscow (1963), 680 pages.

83.* Muchnik, A. A., and Gindikin, S. G., Completeness of a system of unreliable elements realizing a logic-algebraic function, in: Theory of Finite and Probabilistic Automata [in Russian], Nauka, Moscow (1965), pp. 333-335.

84. Nagano Takasi and Oteru Sadamu, Problems in the Synthesis of reliable logic structures from unreliable elements, Trans. Inst. Electron. and Commun. Engrs. Japan, C52(6):335-340 (1969).

85. Nakamiti Matsuro, Problems in the synthesis of reliable logic networks, J. Japan. Assoc. Automat. Control. Engrs., 13(2):116-128 (1969).

86.* Nemsadze, N. K., Encoding of the states of a finite automaton, Probl. Peredachi Inform., 5(1):79-86 (1969).

87.* Nechiporenko, V. I., Structural Analysis and Methods of Synthesis of Reliable Systems [in Russian], Sov. Radio, Moscow (1968), 255 pages.

88.* Nechiporuk, É. I., On self-correcting gating networks, Dokl. Akad. Nauk SSSR, 156(5):1045-1048 (1964).

89.* Nechiporuk, É. I., On topological self-correction principles, Dokl. Akad. Nauk SSSR, 179(4):790-793 (1968).

90.* Nechiporuk, É. I., On topological self-correction principles, in: Problems of Cybernetics [in Russian], No. 21, Nauka, Moscow (1969), pp. 5-102.

91.* Nomokonov, V. N., and Tolstyakov, V. S., Increasing the reliability of binary counters by redundant encoding, Izv. Leningrad Élektrotekh. Inst., No. 56, Part 2 (1966).

92.* Gavrilov, M. A., (ed.), Fundamental Concepts of Automation; Terminology [in Russian], Nauka, Moscow (1966), 19 pages.

93.* Ostianu, V. M., Synthesis of d-reliable relay devices from functional elements and analysis of their reliability (Scientific Council on Cybernetics, Academy of Sciences of the USSR, Kiev House of Scientific-Technical Propaganda, School-Seminar: Methods for the Synthesis of Reliable Systems from Insufficiently Reliable Elements) [in Russian], Kiev (1968), pp. 22-52.

94.* Ostianu, V. M., and Krishtal', V. Z., Synthesis of relay devices insensitive to element failures and input distortions, in: Abstract and Structural Theory of Relay Devices [in Russian], Nauka, Moscow (1966), pp. 176-188.

95.* Ostianu, V. M., Lyakhovich, V. F., and Potekhin, A. I., On the various classes of codes used in computation channels, in: Fourth Conf. Theory of Information Transmission and Encoding, Sec. 5 [in Russian], Moscow-Tashkent (1969), pp. 111-117.

96.* Ostianu, V. M., and Potekhin, A. I., Use of structural redundancy for the assurance of stability and d-reliability of relay devices, Radio Measurements, Materials of Scientific-Technical Conference, 1969 [in Russian], Vilnius (1969), pp. 284-285.

97.* Ostianu, V. M., and Potekhin, A. I., Race-abstract techniques in d-reliable relay devices, Bull. Math. Sci. Math. RS Roumanie, 13(1)(1) (1969).

98.* Ostrovskii, Yu. I., Method for the encoding of transition matrix rows, Avtomat. i Telemekhan., 25:382-393 (1964).

99.* Ostrovskii, Yu. I., Algorithm for the synthesis of multicycle networks in which critical races of the relays are impossible, Avtomat. i Telemekhan., 25(5):844-860 (1965).

100.* Parkhomenko, P.P., and Tomfel'd, Yu. L., Minimization of the number of essential connections between rows of a transition table, in: Structural Theory of Relay Devices [in Russian], Izd. AN SSSR, Moscow (1963), pp. 128-144.

101.* Petri, N. V., Complexity of the realization of logic-algebraic functions by contact networks with unreliable contacts and high reliability demands, in: Problems in Cybernetics [in Russian], No. 21, Nauka, Moscow (1969), pp. 159-169.

102.* Petrosyan, A. V., Aspects of the interference tolerance of logic-algebraic functions, Dokl. Akad, Nauk Arm. SSR, 36(3):147-151 (1963).

103.* Piil', E. I., Method of assignment of internal states of an asynchronous finite automation, in: Information Transmission Problems [in Russian], No. 17, Nauka, Moscow (1964), pp. 56-69.

104.* Piil', E. I., Encoding of internal states of a finite automaton, Izv. AN SSSR, Tekh. Kibernetika, No. 2, pp. 58-65 (1965).

105.* Piil', E. I., Aspects of the encoding of finite automata, in: Problems in the Synthesis of Digital Automata [in Russian], Nauka, Moscow (1967), pp. 14-27.

106.* Peterson, W. W., Error-Correcting Codes, MIT Press, Cambridge, Mass. (1961).

107.* Polovko, A. M., Fundamentals of Reliability Theory [in Russian], Nauka, Moscow (1964), 446 pages.

108.* Potapov, V. I., Reliability of networks using threshold elements that compute universal logic functions, Izv. Leningrad. Élektrotekh. Inst., No. 68, pp. 67-70 (1968).

109.* Potapov, Yu. G., and Yablonskii, S. V., Synthesis of self-correcting contact networks," Dokl. Akad. Nauk SSSR, 134(3):544-547 (1960).

110. Potekhin, A. I., A technique for the transformation of the transition graph of a relay device into a partial subgraph of an n-dimensional unit cube, in: Fifteenth Conf. of Young Specialists, Abstracts of Papers [in Russian], IPU (IAT) (1968).

111.* Potekhin, A. I., Method for the stability and d-reliability assurance of finite automata, All-Republic Seminar: Reliable Synthesis of Digital Automata, Abstracts of Papers [in Russian], Kiev (1969).

112.* Rabinovich, V. M., Self-correcting parity counter networks, in: Problems in Cybernetics [in Russian], No. 17, Nauka, Moscow (1966), pp. 227-231.

113.* Rabinovich, V. M., Catalog of self-correcting contact networks for functions of three variables (open-circuit case), in: Problems in Cybernetics [in Russian], No. 21, Nauka, Moscow (1969), pp. 171-183.

114.* Radchenko, A. N., Relationship fo the theory of error-correcting codes to learning and reliability problems, in: Cybernetics in the Service of Communism [in Russian], Vol. 3, Énergiya, Moscow -Leningrad (1966), pp. 87-114.

115.* Roginskii, V. N., Operation of relay networks in transiet periods, Avtomat. i Telemekhan., 20(10):1409-1416 (1959).

116.* Roginskii, V. N., Synthesis of contact networks with real contacts, Avtomat. i Telemekhan., 22(10):1355-1359 (1961).

117.* Roginskii, V. N., Transient processes in relay devices, Proc. Third All-Union Conf. Automatic Control (Technical Cybernetics), 1965, Self-Adapting Systems [in Russsian], Nauka, Moscow (1967), pp. 313-321.

118.* Roginskii, V. N., Behavior of discrete automata (relay-action devices) in transient periods, in: Theory of Finite and Probabilistic Automata [in Russian], Nauka, Moscow (1965), pp. 136-139.

119.* Roginskii, V. N., Response of a single-cycle discrete automaton to a change of input function, Probl. Peredachi Inform., 1(1):52-56 (1965).

120.* Roginskii, V. N., Transformation of the time parameters of signals in discrete asynchronous automata, Probl. Peredachi Inform., 2(1):100-104 (1966).

121.* Roginskii, V. N., Dynamics of the operation of discrete automata with linear delays, Probl. Peredachi Inform., 3(1):75-78 (1967).

122.* Roginskii, V. N., On the theory of dynamic discrete automata, in: Information Nets and Commutation [in Russian], Nauka, Moscow (1968).

123.* Rudnev, Yu. P., and Khetagurov, Ya. A., Use of linear batched correcting codes in a parallel digital computer, Izv. Vuzov, Élektromekhanika, No. 4, pp. 406-416 (1967).

124.* Ryakin, O. M., Aspects of the checking of combinational networks, in: Problems in the Theory of Electronic Digital Mathematical Machines, Seminar [in Russian], No. 1, Kiev (1967), pp. 85-108.

125.* Ryakin, O. M., Separable codes for the checking of asymmetric errors in combinational networks, Kibernetika, No. 1, pp. 55-58 (1967).

126.* Savchenko, Yu. G., Problems in the synthesis of functionally reliable discrete automata, in: Applied Problems of Technical Cybernetics [in Russian], Sov. Radio, Moscow (1966), pp. 115-122.

127.* Savchenko, Yu. G., Redundant automata using restoring organs with memory, in: Control Computers and Systems [in Russian], Énergiya, Moscow (1967), pp. 160-163.

128.* Savchenko, Yu. G., Error correction in arbitrary digital automata without structural modification (Scientific Council on Cybernetics, Academy of Sciences of the USSR, Kiev House of Scientific-Technical Propaganda, School-Seminar: Methods for the Synthesis of Reliable Systems from Insufficiently Reliable Elements) [in Russian], Kiev (1968), pp. 3-22.

129.* Sagalovich, Yu. L., Methods for increasing the reliability of a finite automaton, Probl. Peredachi Inform., 1(2):27-35 (1965).

130.* Sagalovich, Yu. L., Redundancy methods for increasing the reliability of a finite automaton, Probl. Peredachi Inform., 4(1):62-72 (1968).

131.* Sagalovich, Yu. L., Interference-tolerant encoding of the states of an asynchronous finite automaton, Probl. Peredachi Inform., 2(2):54-59 (1966).

132.* Sagalovich, Yu. L., Complexity of a combinational block for the interference-tolerant encoding of automaton states, Probl. Peredachi Inform., 5(3):37-45 (1969).

133.* Sagalovich, Yu. L., Memory reduction of an automaton resistant to failures and races of its internal elements, Probl. Peredachi Inform., 3(2):73-85 (1967).

134.* Sagalovich, Yu. L., Decoding in automata, in: Fourth Conf. Theory of Information Transmission and Encoding, Sec. 5 [in Russian], Moscow-Tashkent (1969), pp. 135-137.

135.* Sapozhnikov, V. V., and Sapozhnikov, Vl. V., Synthesis of multicycle networks in which critical races of the relays are impossible, Avtomat. i Vychisl. Tekh., No. 5, pp. 9-17 (1968).

136. Safonov, I. V., An algorithm for the design of operational devices with corrective redundancy, All-Union Interinstitutional Conf. Algorithmic Methods for the Design of Digital Systems, Abstracts of Papers [in Russian], Min. Vyssh. i Sr. Spets. Obraz., Leningrad; Inst. Toch. Mekh. i Optiki, Leningrad (1969), 31 pages.

137.* Svechinskii, V. B., Self-correcting designs of finite automata, Avtomat, i Telemekhan., 25(5):685-691 (1964).

138.* Svechinskii, V. B., Discrete automata with multiple lengths, in: Instrument Design, Automatic Devices, and Control Systems [in Russian], Nauka, Moscow (1967), pp. 206-207.

139.* Races in relay systems, Abstracts of Papers, Internat. Sympos. Asynchrony and Absence of Continuity in the Action of Relays and Relay Systems, Bucharest, 1968 [in Russian], Izd. IPU (IAT), Moscow (1969).

140.* USSR State Standard GOST 13377-61: Reliability in Engineering; Terms [in Russian], Moscow (1968).

141. Sugino Kadzue, Inagaki Yasuesi, and Fukumura Teruo, Methods for the detection of races by means of ternary algebra, J. Inst. Electron and Commun. Engrs. Japan, 50(6):997-1004 (1967).

142.* Terminology in the Theory of Relay Devices [in Russian], Inst. Avtomat. i Telemekhan. (Tekh. Kibernetika) (IAT), Moscow (1964), 34 pages.

143.* Artobolevskii, et al. (collective authors), Terminology of fundamental automation concepts, Proc. First Internat. Congr. IFAC [in Russian], Izd. AN SSSR, Moscow (1960), pp. 642-677.

144. Togino Kadzuto, Races in sequential networks, J. Soc. Instrum. and Control Engrs. Japan, 6(12):939-944 (1967).

145.* Tomfel'd, Yu. L., Method of assignment of states of intermediate elements in the synthesis of multicycle relay devices, in: Structural Theory of Relay Devices [in Russian], Izd. AN SSSR, Moscow (1963), pp. 110-127.

146.* Tomfel'd, Yu. L., State-assignment method (review and classification), in: Abstract and Structural Theory of Relay Devices [in Russian], Nauka, Moscow (1966), pp. 26-54.

147.* Tomfel'd, Yu. L., Elimination of races by transformation of the transition graph, in: Abstract and Structural Theory of Relay Devices [in Russian], Nauka, Moscow (1966), pp. 55-79.

148.* Turuta, E. N., On structural redundancy in the synthesis of a C-automaton with specified reliability, in: Information-Transmission Nets and Their Automation [in Russian], Nauka, Moscow (1965).

149.* Turuta, E. N., A method for increasing the reliability of a finite automaton, Probl. Peredachi Inform., 2(4):84-86 (1966).

150.* Turuta, E. N., Use of structural redundancy to increase the reliability of automata, in: Computer Systems [in Russian], No. 25, Nauka, Novosibirsk (1966), pp. 19-30.

151.* Turuta, E. N., Methods for increasing the reliability of finite automata, in: Problems in the Synthesis of Digital Automata [in Russian], Nauka, Moscow (1967), pp. 28-47.

152.* Fabrikant, V. L., Synthesis of reliable networks from unreliable relays, Élektrichestvo, No. 7, pp. 42-48 (1967).

153. Frantsis, T. A., Error correction in asynchronous automata, in: Automatic Control [in Russian], Zinatne, Riga (1967), pp. 87-97.

154.* Grantsis, T. A., Application of correcting codes to increase the reliability of combinational logic networks, in: Theory of Discrete Automata [in Russian], Zinatne, Riga (1967), pp. 201-225.

155.* Frantsis, T. A., and Yanbykh, G. F., Automatic error correction in discrete automata, in: Automatic Control [in Russian], Zinatne, Riga (1967), pp. 53-85.

156.* Frantsis, T. A., and Yanbykh, G. F., Redundancy in Electronic Discrete Devices [in Russian], Énergiya, Leningrad (1969), 248 pages.

157.* Tsertsvadze, G. N., Stochastic automata and problems in the synthesis of reliable automata from unreliable elements (I), Avtomat. i Telemekhan., 25(2):213-226 (1964).

158.* Tsiramua, G. S., A method for the synthesis of complex multifunctional discrete systems with a high level of reliability, in: Control Computers and Systems [in Russian], Énergiya, Moscow (1967), pp. 83-87.

159.* Shannon, C., Papers on Information Theory and Cybernetics [Russian translation], Izd. In. Lit., Moscow (1963), 829 pages.

160.* Shekhovtsov, O. I., Use of code redundancy to increase the reliability of remote-control system encoders, Izv. Leningrad. Élektrotekh. Inst., No. 59, pp. 134-144 (1967).

161.* Yakubaitis, É. A., Asynchronous Logical Automata [in Russian], Zinatne, Riga (1966), 380 pages.

162.* Yakubaitis, É. A., Method of encoding of internal states of a finite automaton, Avtomat. i Vychisl. Tekh. No. 3, pp. 1-3 (1968).

163.* Yakubaitis, É. A., Gobzemis, A. Yu. Gorobets, V. G., and Fritsnovich, G. F., Minimization of the memory volume of sequential asynchronous logic networks, in: Automation and Computer Engineering [in Russian], No. 12, Zinatne, Riga (1966), pp. 115-130.

164. Yakubaitis, É. A., and Gorobets, V. G., Synthesis of sequential asynchronous logical automata on the basis of three types of modules, in: Automatic Control [in Russian], Zinatne, Riga (1967), pp. 25-37.

165.* Amarel, S., and Brzozowski, J. A., Theoretical considerations on reliability properties of recursive triangular switching networks, in: Redundancy Techniques for Computing Systems, Spartan Books, Washington, D. C. (1960), pp. 70-128.

166.* Angell, J. V., The need and means for self-repairing circuits, IEEE Internat. Conv. Rec., 11(2):193-199 (1963).

167.* Armstrong, D. B., A general method of applying error-correction to synchronous digital systems, Bell System Tech. J., 40(2):577-593 (1961).

168.* Armstrong, D. B., A programmed algorithm for assigning internal codes to sequential machines, IRE Trans. Electronic Computers, EC-11(4):466-472 (1962).

169.* Armstrong, D. B., On the efficient assignment of internal codes to sequential machines, IRE Trans. Electronic Computers, EC-11(5):611-622 (1962).

170.* Armstrong, D. B., Friedman, A. D., and Menon, P. R., Synthesis of asynchronous sequential circuits with minimum number of delay elements, IEEE Conf. Rec. 8th Annual Asympos. Switching and Automata Theory, Austin, Texas, 1967, New York (1967), pp. 95-105.

171.* Barlow, R. E., Hunter, L. C., and Proschan, F., Optimum redundancy when components are subject to two kinds of failure, J. Soc. Indust. Appl. Math., 11(1):64-73 (1963).

172.* Barlow, R. E., and Proschan, F., Mathematical Theory of Reliability, Wiley, New York (1965), xiii + 256 pages.

173. Bazovsky, I., Reliability Theory and Practice, Prentice-Hall, New York (1961).

174.* Beister, J., and Görke, W., Ausfallsicherung bei sequentiellen Schaltungen, Elektron. Rechenanlag., 10(3):123-130 (1968).

175. Bellomi, C., Le reti difettive nell'algebra dei circuiti variabili, Ingegneria Ferroviaria, 16(10):907-922 (1961).

176. Bidoul, J., Étude des possibilités d'augmentation de la fiabilité des machines logiques et numériques, Revue, A, 8(3):112-122 (1966).

177.* Blake, D. V., Open and short circuit failure of hammock networks, Electron. Reliability and Microminiatur. (December, 1963).

178.* Bläss, B., Überlegungen zur Zuverlässigkeit von Sicherheitssystemen, Regelungstechnik, 17(6):266-270 (1969).

179. Bojarski, W. W., Wsprawie artykulu p^t, "Teoria niezawodnósci i jej zastosowanie w Polsce w latach 1957-199," J. Migadskego, Przegl. Elektrotekh., 44(1):40-41 (1968).

180.* Boshko, O., Glinski, G. S., and Therrien, J., Reliable networks from unreliable components, Trans. Engng. Inst. Canda, 5(3):184-201 (1961).

181.* Brown, W., Tierney. J., and Wasserman, R., Redundancy improves computer reliability, in: Redundancy Techniques for Computing Systems, Spartan Books, Washington, D. C. (1962), 378 pages.

182.* Buzzell, G., Nutting, W., and Wasserman, R., Majority gate logic improves digital system reliability, IRE Internat. Conv. Rec., 9(2):264-271 (1961).

183.* Calabro, S. R., Reliability Principles and Practices, McGraw-Hill, New York (1962), 371 pages.

184.* Cluley, J. C., Low-level redundancy as a means of improving digital computer reliability, Electron. Reliability and Microminiatur., No. 1, pp. 203-216 (July-September, 1962).

185. Cluley, J. C., The reliability of electronic systems, Radio Electron. Engr., page 110 (1966).

186. Cohen, A., Reliability in complex systems, Twelfth Ann. Tech. Conf., Trans. Amer. Soc. Quality Control, New York (1966); Milwaukee (1966), pp. 667-675.

187.* Coroi-Nedelcu, M., Hazard and race phenomena in switching circuits with state devices, Abstracts of Papers, Internat. Sympos. Asynchrony and Absence of Continuity in the Action of Relays and Relay Systems, Burcharest, 1968, Izd. IPU (IAT), Moscow (1970).

188.* Cowan, J. D., Many-valued logics and reliable automata, in: Principles of Self-Organization, Pergamon, New York (1962), pp. 135-179.

189. Cowan, J. D., Synthesis of reliable automata from unreliable components, in: Automata Theory, Academic Press, New York (1966), pp. 131-145.

190.* Deo, N., Generalized parallel redundancy in digital computers, IEEE Trans. Electronic Computers, EC-17(6):600(1968).

191.* dePian, L., and Grisamore, N. T., Reliability using redundancy concepts, IRE Trans. Reliability and Quality Control, RQC-9:53-60 (1960).

192.* Dolotta, T. A., and McCluskey, E. J., The coding of internal states of sequential circuits, IEEE Trans. Electronic Computers, EC-13(5):549-562 (1964).

193. Dunning, M., and Kolman, B., Reliability and fault-masking in n-variable NOR trees, IEEE Conf. Rec. Switching Circuit Theory and Logical Design, Ann Arbor, Michigan, 1965, IEEE, New York (1965), pp. 126-142.

194.* Eden, M., A note on error detection in noisy logical computers, Information and Control, 2(3):310-313 (1959).

195.* Eichelberger, E. B., Hazard detection in combinational and sequential switching circuits, IEEE Conf. Rec. Switching Circuit Theory and Logical Design, Ann. IEEE, New York (1964), pp. 111-120.

196.* Elias, P., Computation in the presence of noise, IBM J. Res. Develop., 2(4):346-353 (1958).

197.* Esary, J. D., and Proschan, F., The reliability of coherent systems, in: Redundancy Techniques for Computing Systems, Spartan Books, Washington, D.C. (1962), pp. 47-61.

198.* Fan Liang Tseng, Wang Chiu Sen, Tillman, F. A., and Hwang Ching Lai, Optimization of system reliability, IEEE Trans. Reliability, R-16(2):81-86 (1967).

199.* Farrell, E. J., Improving the reliability of digital devices with redundancy, an application of decision theory, IRE Trans. Reliability and Quality Control, RQC-11(1):44-52 (1962).

200.* Fleck, J. J., Redundancy techniques for reliable flight-control computer, IEEE Trans. Communication and Electronics, No. 68, pp. 535-546 (1963).

201. Francois, E., Analyse et synthèse de structures logiques selon le critère de fiabilité, une méthode de synthèse particulière aux logiques sequentiells, Ann. Radioelectron., 20(84):323-340 (1965).

202.* Frank, H., and Yau, S. S., Improving reliability of a sequential machine by error-correcting state assignments, IEEE Trans. Electronic Computers, EC-15(1):111-113 (1966).

203. Friedman, A. D., Feedback in asynchronous sequential circuits, IEEE Trans. Electronic Computers, EC-15(5):740-749 (1966).

204.* Friedman, A. D., A decision procedure for computations of finite automata, J. Assoc. Compt. Mach., No. 9, pp. 315-323 (1962).

205. Gabor, A., Adaptive coding for self-clocking recording, IEEE Trans. Electronic Computers, EC-16(6):866-868 (1967).

206.* Caldwell, S. H., Switching Circuits and Logical Design, Wiley, New York (1958).

207. Galimberti, R., Morpurogo, R., Sue le codage de variables internes des circuits sequentielles asynchrones, Automatisme, 13(4):160-163 (1968).

208.* Gavrilov, M. A., Ostinau, V. M., Krishtal', V. Z., The synthesis of reliable relay networks insensitive to (d_0 d_1) asymmetrical failures of internal elements and inputs, Third IFAC Congr., Vol. 1, Book 2, Inst. Mech. Engnrs, London (1968), 833 pages.

209.* Goldman, H. D., Protecting core memory circuits with error-correcting cyclic codes, IEEE Trans. Electronic Computers, EC-13(3):303-304 (1964).

210. Goldman, H. D., Circuit failure asymmetrics for reliability improvement in digital circuits, IEEE Internat. Conv. Rec., 9:115 (1966).

211.* Gore, W., System redundancy and information theory, in: Redundancy Techniques for computing systems, Spartan Books, Washington, D. C. (1962), pp. 294-303.

212.* Görke, W., Selbstkorrigierende Decodierschaltungen fur Hamming-Codes mit Mindestabstanden zwei, drei und vier, NTZ, 19(6):352-367 (1966).

213. Grenander, U., Can we look inside an unreliable automaton? in: Research Papers on Statistics, Wiley, New York (1966), pp. 107-123.

214.* Griesmer, J. H., Miller, R. E., Roth, J. P., and Thomas, J., The design of digital circuits to eliminate catastrophic failures, in: Redundancy Techniques for Computing Systems, Spartan Books, Washington, D. C. (1962), pp. 328-348.

215.* Hall, K. L., Basic rules for designing reliability into semiconductor circuits, Electronics, 36(15):62-66 (1963).

216.* Harrison, M. A., On the error-correcting capacity of finite automata, Information and Control, 8(4):430-460 (1965).

217.* Hartmanis, J., On the state assignment problem for sequential machines (I), IRE Trans. Electronic Computers, EC-10(2):157-165 (1961).

218.* Hartmanis, J., and Stearns, R. E., Some dangers in state reduction of sequential machines, Information and Control, 5(3):252-260 (1962).

219.* Hartmanis, J., and Stearns, R. E., A study of feedback and errors in sequential machines, IEEE Trans. Electronic Computers, EC-12(3):223-232 (1963).

220.* Hazeltine, B., Encoding of asynchronous sequential circuits, IEEE Trans Electronic Computers, EC-14(5):727-729 (1965).

221.* Hlavička, J., Race-free assignment in asynchronous switching circuits, Inform. Process. Mach., No. 13, pp. 99-112 (1967).

222. Howe, A. B., and Coates, C. L., Logic hazards in threshold networks, IEEE Trans. Electronic Computers, EC-17(3):238-251 (1968).

223.* Huffman, D. A., The synthesis of sequential switching circuits, J. Franklin Inst., 257(3):4(1954).

224.* Huffman, D. A., A study of the Memory Requirements of Sequential Switching Circuits, MIT Res. Lab. Electronics, Tech. Rep. (1955), 293 pages.

225.* Huffman, D. A., The design and use of hazard-free switching networks, J. Assoc. Comput. Mach., 4(1):47-62 (1957).

226.* Ichikawa Tadao and Watanabe Teruji, Failure-operable redundant NAND networks, Abstr. Trans. JIECE Japan, 52(8):23-25 (1969).

227.* Inskip, F. A., Redundancy in digital systems., Electron. Engng., 39(470): 244-249, 276, 277 (1967).

228.* Jensen, P. A., Quadded NOR logic, IEEE Trans. Reliability, R-12(3):22-31 (1963).

229.* Jensen, P. A., The reliability of redundant multiple-line networks, IEEE Trans. Reliability, R-13(1):23-33 (1964).

230.* Kämmerer, W., Zur Konstruktion selbstkorrigiernder Automaten, Elektronische Informationsverarbeitung und Kybernetik, 2(3):165-175 (1966).

231.* Kämmerer, W., Zur Struktur selbstkorrigierender Automaten, Messen. Steuern Regeln., 1(1):5-13 (1967).

232.* Kautz, W., Codes and coding circuitry for automatic error correction within digital systems, in: Redundancy Techniques for Computing Systems, Spartan Books, Washington, D. C. (1962), pp. 151-195.

233.* Kemp, J. C., Redundant digital systems, in: Redundancy Techniques for Computing Systems, Spartan Books, Washington, D. C. (1962), pp. 285-293.

234. Kenvaras, N., and Lagoyinnis, D., Reduction of hazard in sequential circuits, Electron. Engng. (GB), 40(487):524-528 (1968).

235.* Klaschka, T. Optimal use of partial redundancy and the system reliability which results, Electron. Lett., No. 1 (1968).

236. Klega, V., O jednom stochastickém modelu spolehlivosti, Aplikace Mat., 11(3):224-231 (1966).

237. Klimal, M. L., Asynchronous electronic switching circuits, IRE Nat. Conv. Rec., No. 4, pp. 267-274 (1959).

238.* Kochen, M., Extension of Moore—Shannon model for relay circuits, IBM J. Res. Develop., 3(2):169-186 (1959).

239. Lawrence, L. A. J., High-security control through redundant channels, Control. Engng., 12(3):74-79 (1965).

240.* Levy, S., The reliability of recursive triangular switching networks built of rectifier gates, in: Redundancy Techniques for Computing Systems, Spartan Books, Washington, D. C. (1962), pp. 129-151.

241. Lin Wen-chen, Asynchronous sequential machine with a predetermined degree of reliability, Acta Automatica Sinica, 4(1):45 (1966).

242.* Lindman, R., A theorem for deriving majority-logic networks within an augmented Boolean algebra, IRE Trans. Electronic Computers, EC-9:338 (1960).

243.* Liu, C. N., A state variable assignment method for asynchronous sequential switching circuits, J. Assoc. Comput. Mach., 10(2):209-216 (1963).

244.* Liu, Chung Laung, and Liu, Jane W. S., On a multiplexing scheme for threshold logical elements, Information and Control, 8(3) (1965).

245.* Löfgren, L., Automata of high complexity and methods of increasing their reliability by redundancy, Congr. Internat. Automat., Paris, 1956, Paris-Brussels (1959), pp. 34-42.

246.* Lowenschuss, O., Restoring organs in redundant automata, Information and Control 2(2):113-136 (1959).

247.* Lowrie, R. W., High-reliability computers using duplex redundancy, Electron. Indust., 22(8):116 (1963).

248.* Lyons, R. E., and Vanderkulk, W., The use of triple-modular redundancy to improve computer reliability, IBM J. Res. Develop. 6(2):200-209 (1962).

249. Maitra, K. K., Stability of logical networks and its application to improvement
      of reliability, IRE Trans. Circuit Theory, CT-9(3):335-341 (1961).

250.* Mann, W. C.,Systematically introduced redundancy in logical systems, IRE
      Internat. Conv. Rec., 2:241-263 (1961).

251.* Mann, W. C., Restorative processes for redundant computing systems, in:
      Redundancy Techniques for Computing Systems, Spartan Books, Washington,
      D. C. (1962), pp. 262-284.

252. Marianowicz, T. P., Niektóre metody matematyczne teorii niezawodnósci,
      Przegl. Elektrotech., 41(5):173-178 (1965).

253. Marionov, P. N., and Page, E. W., The organization of a self-repairing system
      from multifunctional logic elements, Proc. IEEE, 57(7):1320 (1969).

254. Maxwell, L. M., Synthesis of contact networks from prescribed reliability
      functions, J. Franklin Inst., 281(3):214-234 (1966).

255.* McCulloch, W. S., and Pitts, W. H., A logical calculus of the ideas immanent
      in nervous activity, in: Information Storage and Neutral Control, Thomas,
      Springfield, Ill. (1963), pp. 379-399.

256. McGhee, R. B., Some aids to the detection of hazards in combinational
      switching circuits, IEEE Trans. Computers, C-18(6):561-565 (1969).

257. Meisel, W. S., and Collins, D. C., Hazards in noncritical races, Proc. IEEE,
      56(8):1361-1363 (1968).

258. Migdalski, J., Teoria niezawodnósci i jej zastosowania w latach 1957-1966,
      Przegl. Elektrotekh., 43(5):189-190 (1967).

259. Miller, R. E., Switching Theory, Vol. I: Combinational Circuits, Wiley,
      New York (1965), xiii + 351 pages.

260.* Miller, R. E., Switching Theory, Vol. II: Sequential Circuits and Machines,
      Wiley, New York (1965), 250 pages.

261.* Mine Hisashi and Koga Yoshiaku, Basic properties and a construction method
      for fail-safe logical systems. IEEE Trans. Electronic Computers, EC-16(3):
      282-289 (1967).

262. Moisil, G. C., Sur les aleas de continuité dans les circuits de commutation,
      automatisme, 10(9):313-315 (1965).

263.* Moore, E. F., and Shannon, C. E., Reliable circuits using less reliable relays,
      Part I, J. Franklin Inst., 262(3):191-208 (1956).

264. Moore, E. F., and Shannon, C. E., Reliable circuits using less reliable relays,
      Part II, J. Franklin Inst., 262(4):281-297 (1956).

265. Muller, D. E., and Bartky, W. S., A theory of asynchronous circuits, Proc.
      Sympos. Switching Theory, Part 1, Harvard Univ. Press, Cambridge, Mass.
      (1959), pp. 204-243.

266.* Mullin, A. A., On the nature of the reliability of automata, in: Redundancy
      Techniques for Computing Systems, Spartan Books, Washington, D. C. (1962),
      pp. 196-204.

267. Muroga, S., Functional forms of dual-comparable functions and a necessary
      and sufficient condition for realizability of a majority function, IEEE Trans
      Communication and Electronics, CE-83(74):474-486 (1964).

268.* Nadler, M., Some questions of computer reliability through redundancy,
      Stroje Zpracov. Inform., No. 7 (1960).

269.* Neumann, J. von., Probabilistic logics and the synthesis of reliable organisms
      from unreliable components, in: Automata Studies, Ann. Math. Studies,
      No. 34, Princeton Univ. Press, N. J. (1956), pp. 43-98.
270.* Ord-Smith, R. J., An extension of block design methods and an application in
      the construction of redundant fault-reducing circuits for computers, Comput.
      J., 8(1):28-32 (1965).
271.* Ostianu [Ostianou], V. M., Étude des possibilités d'augmentation de la fiabilité
      des systemes logiques, in: Systemes Logiques, Conception et applications,
      Coll. Internat., Sept. 15-20, 1969, Vol. 2, AICA, Brussels (1969), pp. 703-811.
272.* Palounek, L., Návrh spolehlivého zařízeníz méne spolehlivých součástí Slabo-
      proudý Obzor, 24(1):664-670 (1963).
273.* Peterson, W. W., and Rabin, M. O., On codes for checking logical operations,
      IBM J. Res. Develop., 3(2):163-168 (1959).
274.* Pian, L. de., and Grisamore, N. T., Two approaches to incorporating redund-
      ancy into logical design, in: Redundancy Techniques for Computing Systems,
      Spartan Books, Washington, D. C. (1962), pp. 379-388.
275.* Pierce, W. H., Adaptive vote-takers improves the use of redundancy, in:
      Redundancy Techniques for Computing Systems, Spartan Books, Washington,
      D. C. (1962), pp. 229-250.
276.* Pierce, W. H., Adaptive decision elements to improve the reliability of
      redundancy systems, IRE Internat. Conv. Rec., 10(4):124-131 (1962).
277.* Pierce, W. H., Interwoven redundant logic, J. Franklin Inst., 277(1):55-85
      (1964).
278.* Pierce, W. H., Failure-Tolerant Computer Design, Academic Press, New York
      (1965).
279. Pierce, W. H., Asymptotic properties of systems synthesized for maximum relia-
      bility, Information and Control, 7(3):340-359 (1964).
280.* Polhemus, J. T., Magnetic code command decoders capable of error correction,
      Rec. Internat. Space Electronics Sympos., Las Vegas, 1964, 2d-1-2d-6.
281.* Rao Thammavarapu, R. N., Use of error-correcting codes on memory words
      for improved reliability, Trans. IEEE, 17(2) (1968).
282.* Rau, J. G., Redundancy and trichotomous systems, J. Soc. Indust. Appl. Math.,
      12(4):827-837 (1964).
283.* Repton, C. S., Estimating the optimum position for restoring organs in noncas-
      caded redundant networks, Microelectronics and Reliability, 8(1):23-31 (1969).
284.* Reza, R., A note on reliability functions, Proc. Second Internat. Congr. Cy-
      bernetics, Assoc. Internat. Cybernetique, Namur (1960), 509-520.
285.* Roberts, D. C., Increasing reliability of digital computers, Computer Design,
      8(1):44-48 (1969).
286.* Roy-Chaudhuri, D. K., On the construction of minimally redundant reliable
      system designs, Bell Syst. Tech. J., 40(2):595-611 (1961).
287.* Russo, R. L., Synthesis of error-tolerant counters using minimum-distance
      three-state assignments, IEEE Trans. Electronic Computers, EC-14(3):359-366
      (1965).
288.* Saucier, G., Encoding of asynchronous sequential networks, IEEE Trans.
      Electronic Computers, EC-16(3):365-369 (1967).

289.* Saucier, G., Algorithmes de codage des automates asynchrones, Automatisme, 14(10):520-523 (1969).

290. Sethares, G., Closed sets of Boolean functions and the reliability problem for polyfunctional nets, IEEE Trans. Electronic Computers, EC-15(1):115-117 (1966).

291.* Akers, Sheldon B., Jr., A diagrammatic approach to multilevel logic synthesis, IEEE Trans. Electronic Computers, EC-14(5) (1965).

292. Shirakawa Isao, Kasami Tadao, and Ozaki Huroshi, On the coding of the paths in a graph and its applications, Tech. Rep. Osaka Univ., 16 (March, 1966).

293. Stanciulesco, F., La concéption et la réalisation électronique des schémas logiques à grande fiabilité, Automatisme, 11(11):597-602 (1966).

294.* Stearns, R. E., and Hartmanis, J., On the state assignment problem for sequential machines (II), IRE Trans. Electronic Computers, EC-10(4):593-603(1961).

295.* Swoboda, J., Binäre Gruppencodes zur sicherung logischer Schaltkreise gegen Fehler, Elektron. Rechnanlag., 7(2):85-90 (1965).

296.* Teoste, R., Design of a repairable redundant computer, IRE Trans. Electronic Computers, EC-11(5):643-649 (1962).

297.* Teoste, R., Digital circuit redundancy, IEEE Trans. Reliability, R-13(2):42-61 (1964).

298.* Tooley, J., Network coding for reliability, IEEE Trans. Communication and Electronics.

299.* Tracey, J. H., Internal state assignments for asynchronous sequential machines, IEEE Trans. Electronic Computers, EC-15(4):551-560 (1966).

300.* Tryon, J. G., Quadded logic, in: Redundancy Techniques for Computing Systems, Spartan Books, Washington, D. C. (1962), pp. 205-228.

301.* Unger, S. H., Hazards and delays in asynchronous sequential switching circuits (I), IRE Trans. Circuit Theory, CT-6(1) (1959).

302.* Unger, S. H., Hazards and delays in asynchronous sequential switching circuits (II), IRE Trans. Circuit Theory, CT-6(1) (1959).

303. Unger, S. H., A row assignment for delay-free realizations of flow tables without essential hazards, IEEE Trans. Computers, C-17(2):146-151 (1968).

304.* Urbano, R. H., On the convergence and ultimate reliability of iterated neural nets, IEEE Trans. Electronic Computers, EC-13(3):204-225 (1964).

305.* Verbeek, L. A. M., Reliable computation with unreliable circuitry, MIT Tech. Res. Lab. Electron. (1960).

306. Verbeek, L. A. M., Notes on the synthesis of infallible networks, Quart. Progr. Rep. MIT (Cambridge, Mass.), 15(61):187-191 (1961).

307.* Wallmark, J. T., and Revest, A. G., Redundancy in unipolar field-effect transistor circuits, IEEE Trans. Electronic Computers, 12(1):23-25, 288 (1963).

308.* Winograd, S., Coding for logical operations, IBM J. Res. Develop., 6(4):430-436 (1962).

309.* Winograd, S., Redundancy and complexity of logical elements, Information and Control, 6(3):177-194 (1963).

310.* Winograd, S., Input-error-limiting automata, J. Assoc. Comput. Mach., 11(3):338-351 (1964).

311.[*] Winograd, S., and Cowan, J. D., Reliable Computation in the Presence of
Noise, MIT Press, Cambridge, Mass. (1963), xiv + 96 pages.

312.[*] Youngblood, J. L., and Breipohl, A. M., Distributed redundancy in two-layer
threshold logic networks, IEEE Trans. Reliability, R-18(1):15-20 (1969).

313.[*] Zandeh, F., Selbstkorrigierende Zuordner für vollständige Codes, Kybernetik,
2(3):114-124 (1964).